Lunar Explorations - The Moon In Plain English

Laurence Waite

Published by Alison Waite, 2024.

LUNAR EXPLORATIONS - THE MOON IN PLAIN ENGLISH

First edition. August 12, 2024.

ISBN: 979-8227361424

Written by Laurence Waite.

Table of Contents

DEDICATION

In memory of my father. As I gaze upon the stars, I find solace in knowing he watches over me.

Laurence Edward Waite 1937-2014

INTRODUCTION

Life on Earth is impacted daily by the Moon in a mysterious yet fascinating fashion as it toils with tidal forces, guides migration, influences our weather and much more. Its moonlight can trigger species' procreation, coral spawning on the Great Barrier Reef, and perhaps even affect human behaviour.

The Author, an amateur astronomer and photographer, provides a unique and interesting perspective on the Moon through facts and detailed imagery. As readers journey through the pages, they will find themselves questioning and indeed pondering some of the key components of the Moon's existence, its creation, and its necessity for Earth's survival.

Through the Author's research and observations, we look at the lunar surface's enigmatic landscape and examine the effects the Moon has on our oceans, seasons, and imagination.

From age-old tales and myths to modern festivals, music and movies, the spell cast by our closest celestial neighbour is indeed a potent one.

Engage in various concepts such as future lunar and space exploration, the necessary resources that may already be available on the Moon and its foreseeable colonisation. Photographs displaying the lunar Alps, craters, and rilles taken with modern telescopes, lunar missions, and satellites are used to demonstrate the wondrous detail of the lunar surface.

Whether you're interested in science and looking to learn more or just curious about the Moon, this book is a valuable resource for those eager to explore its wonders in simple language.

"Lunar Explorations – The Moon in Plain English" simplifies intricate ideas for all readers with its engaging writing style and clear explanations.

By turning the page, you will begin your exploration and although you will find each chapter ties in with the next, feel free to skip ahead and be guided by your curiosity. The path you choose is uniquely yours. Whether you follow the ordered path or meander through the content, the destination remains the same, a deeper understanding and appreciation of our moon.

"That's one small step for man. One giant leap for mankind" Neil Armstrong (1969*).*
-Unsplash/NASA

A close-up view of the laser ranging retroreflector (LR3) which the Apollo 14 astronauts deployed on the Moon during their lunar surface extra vehicular activity (EVA). – NASA

Laser Ranging Retroreflector, this is one of 5 in total, the first being installed by the Apollo 11 crew on July 21, 1969. By shooting a laser from Earth to the Moon and reflecting it, we can determine the current distance between the Moon and Earth.

MEET OUR CELESTIAL NEIGHBOUR

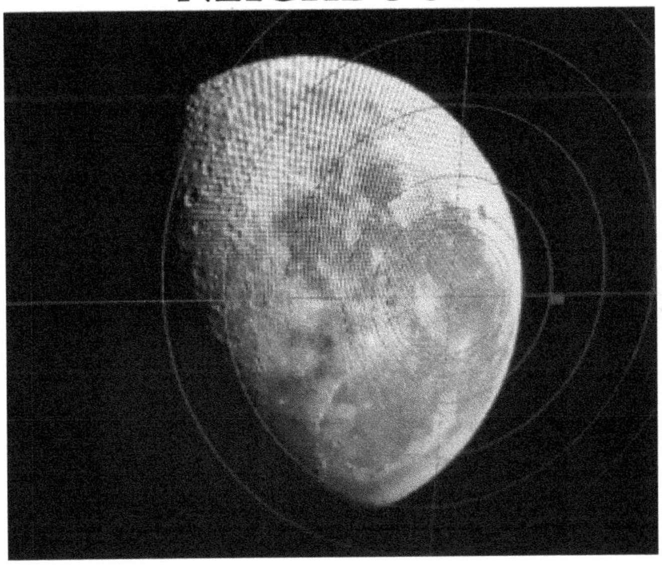

Humans have always held the Moon in high regard on account of its integral part in our history and culture. Throughout different civilisations, it has been admired and celebrated for its mysterious nature and stunning overarching presence. Some cultures worshipped and even offered sacrifices to the Moon in the belief that the lunar deity would repay their loyalty in some way.

This beneficial relationship, as with most worshippers of the gods, was thought to bring luck and protection in times of danger, ensuring strength in battle, or praying for a bountiful harvest.

Traditional practices and beliefs associated with the Moon have played a significant role in defining the spiritual customs of countless societies, connecting them with some distant world beyond.

The Moon along with its shining beams of light, has also deeply impacted artistic and literary works. Stretching from ancient messages conveyed through verses and myth to contemporary tales and melodies, the Moon has been an inspiration unleashing human creativity to no bounds. Its glowing existence in the nocturnal sky has led to the creation of endless artistic masterpieces, and has always been a symbol of romance, mystery, and wonder!

The awareness of the cultural significance of the Moon allows us to appreciate its profound impact on humanity's collective consciousness. It reveals the universal fascination with our only natural orbiting satellite, transcending time and bridging cultures across the globe.

We will attempt to peek into the secrets of the Moon's origin, composition, and the charismatic allure it has on humanity. We also will discover the science behind lunar tides, phases, and eclipses, leaving no stone unturned to reveal the influence of these occurrences on our environment and the perception of our species. With so much to explore, the Moon surely is more than just a satellite!

To fully appreciate the significance of the Moon, we should go over a few basics. So, let us begin with some fundamental, if not interesting, moon statistics.

- At about 1/6th the size of Earth, the Moon has a diameter of about 3,474 km (2,159 miles).
- It orbits Earth at approximately 384,400 km (238,900 miles) from us.
- The Moon has a radius of 1,740 km (1,080 miles).
- The Moon is moving about 3.5cm, (1 ½ inches) further away from Earth, each year. It would take over a billion years

before it would have any major implications for Earth.

- The Moon has an exosphere, not an atmosphere. The exosphere is so thin that it does not offer any protection from radiation emitted by the Sun.
- The Moon's exosphere contains mainly hydrogen, neon, and argon.
- Mostly composed of rock and metal, the Moon's composition is not dissimilar to Earth. However, there are a few differences.
- The Moon's atmosphere is much too thin to generate weather as we know it here on Earth.
- Burning up with heat, we're looking at roughly 121 degrees Celsius (250F) during a moon day and freezing down to -133 Celsius at night. NASA's Lunar Reconnaissance Orbiter (LRO) has located a drop of -246 degrees Celsius in some deep craters that appeared to contain ice.

"On the Moon, snow does not fall. Thunder never rolls. No clouds form in the pitch-black sky." *- someone in NASA.*

That just about covers the weather on the Moon, extreme compared to Earth!

In understanding lunar exploration and discoveries, I think it is also prudent that the reader is somewhat familiar with some key features of the Moon. Some of those features include a lunar phase.

What is a lunar phase you may ask?

A lunar phase refers to the different appearances of the Moon, as it orbits around Earth, sort of. You have probably noticed changes in the Moon's shape that vary slightly from one night to the next.

As we know, the Moon itself is not changing shape, but rather the sunlight shining on a particular part of the Moon at the time reflects in such a way to be visible from Earth. More on lunar phases later.

Looking at the Moon's surface, we can see an assortment of distinct features. Features like the maria "seas", (*plural*) or mare (*singular*). Despite the name, these seas are barren wastelands that probably never contained water. Maria are large dark areas on the Moon's surface created by ancient fiery volcanic activities. On the other side, you have the highlands. Bright and mountainous regions that are composed of older rock formations.

Another noticeable feature on the Moon's surface is its thousands of craters. These craters are formed by the bombardment of asteroids and meteoroids over a period stemming millions of years. With no wind or weather on the Moon, there is little erosion so craters on the Moon, unlike most on Earth, (of which we have about 180), are there forever.

For Earth, one of the crucial roles of the Moon is its influence on our tides with its dominating, yet essential, gravitational force. With its well-rehearsed orbit around our planet, the Moon delivers a gravitational force causing bulges in Earth's oceans.

The effect of these bulges creates our high and low tides. Tides are one of the most understood effects that the Moon has on our planet. Tides do not just influence coastal areas; they also play an indispensable role in our marine life and biodiversity.

In addition, the Moon's gravitational force stabilises Earth's rotational axis, helping to maintain a relatively stable climate and seasonal patterns. Just imagine Earth without the Moon's stabilising effect.

Without the Moon's magical gravitational spell over us, our axis would undergo significant fluctuations. This would cause unpredictable shifts in climate and weather patterns. You see, the Moon acts a lot like a gravitational anchor. It ensures a stable and habitable environment on our planet. Without it, the effects on our planet would be chaotic, to say the least.

Our nights would become darker, and the rotation of Earth may speed up creating days that would last 12 or even 6 hours, maybe less! Our ocean tides would shrink, and our axial tilt would shift affecting weather and seasons. There would be dramatic increases in wind speeds and we would have no more eclipses, of any kind. No doubt about it, there would be significant changes to our planet, none of which I think would be for the good. Take a moment to reflect on the consequences of no moon.

The relationship between the Moon and Earth is extremely interesting and continuous research and understanding is vital in our explorations and future space travel. We are so lucky, as the Moon's proximity to Earth makes it a prime subject for study and exploration by scientists and amateurs alike.

The Moon throughout our history has played a very important role in not only influencing our climate and weather patterns but in ensuring our survival.

On the previous page, we touched on the gravitational effect we experience from the Moon that influences our wind and weather patterns. This results in variations in weather conditions across different vast regions.

This comprehension of the Moon's relationship with Earth is not only important for scientific exploration but also has practical implications.

For example, accurate tidal information is considered crucial for modern maritime activities.

The understanding of the Moon's impact on climate and weather provides us with more accurate forecasting models and enables better preparedness for extreme weather events. The influences the Moon has over Earth, such as the tides, rotational stability, and climate patterns, provide scientists with valuable insights into the dynamics of other planets, their moons, and their interactions. This knowledge further contributes to our understanding of planetary formation, evolution, and the broader field of astrophysics.

The Moon's orbital path is often depicted in illustrations as a perfect circle. This is not the case. The Moon orbits Earth in an anticlockwise, elliptical path.

With this being the actual case, then at certain points within its orbit, the Moon is closer to Earth than at others (perigee).

This elliptical orbital shape contributes to various phenomena we observe from Earth, such as the changing appearance of the Moon and the occurrence of lunar eclipses, but more on that later.

Another seemingly fascinating lunar characteristic is the synchronous rotation with Earth. A synchronous orbit is considered rare but certainly not unheard of. There are about 20 other moons in our solar system alone that share this feature with their host planet.

This movement of the Moon means that it takes about the same amount of time to complete one orbit around Earth as it does to rotate once on its axis. Therefore, we always see the same face of the Moon from Earth. This is what is called 'synchronous tidal locking' and is quite common with larger moons in our solar system.

The whole concept of lunar phases is closely tied to the Moon's orbit. When the Moon orbits around Earth, different regions of its surface get sunlight exposure. The illuminated part of the Moon then becomes visible to us on Earth, resulting in nightly changes of light to various areas on the Moon known as lunar phases. These changes occur before the Moon finishes its month-long orbital cycle and starts anew each time.

When the Moon is positioned between the Earth and the Sun, with the far side of the Moon facing the Sun, the entire far side gets fully illuminated and the near side, (not receiving any sunlight) is not visible to Earth. This indicates a distinct stage of the Moon's orbit phase, a 'new moon'.

In addition, arising from the Moon's orbit, are lunar eclipses, which are often intriguing spectacles.

When the Earth aligns between the Sun and Moon, casting a shadow over the Moon and obscuring its brightness, this is known as a lunar eclipse.

Lunar eclipses can be partial or total, contingent on the Moon's depth in Earth's shadow. These extraordinary events grant a unique chance to observe the Moon's relationship with Earth and its position in the solar system. More on lunar eclipses later.

The lunar month's significance lies in the Moon's celestial positioning during the night. This period represents the cycle from one new moon to the next, lasting about 29.5 days, slightly varying from a standard calendar month. This discrepancy stems from our calendar which is based on the solar system's synodic alignment, while the lunar month aligns with the Moon's orbital rhythm in the sidereal system.

Understanding the Moon's orbit intricacies aids in comprehending phenomena like phase transitions and eclipses. Elements like the

Moon's elliptical pathway, Earth's synchronous rotation, and the concept of a lunar month all influence its dynamic presence in the nocturnal sky. Observing this continuous process allows us to expand our knowledge regarding general orbital characteristics. It also grants us a deeper admiration for our moon and its contribution to shaping our understanding of the universe. It repeatedly emphasizes the importance of ongoing moon research and its significant scientific value in our pursuit to comprehend planet formation and evolution.

Through lunar exploration, scientists have acquired some major insights and unearthed valuable data concerning the Moon's composition, structure, and geological activities. We have gathered information from lunar missions that has also played a crucial part in driving space exploration and technology forward. For example, the Apollo missions enabled scientists to test new technologies and methods that paved the way for upcoming space expeditions. This technological progress achieved during the Apollo era continues to influence diverse fields like robotics, analytics, and communication.

Apogee and Perigee

As stated earlier, the Moon's elliptical orbit means that at some stage each month the Moon will be closer to the Earth than usual. Alternatively, at one stage it will also be further away.

The point at which the Moon is closer is called perigee and appears about 13-15% bigger and about 26-30% brighter.

When a full moon occurs close to the perigee, we call this a Supermoon.

Astrologists believe that a Supermoon increases your emotional awareness, whilst others believe it brings good luck. Whereas astronomers, the science guys, explain it as a straight-line configuration of three celestial bodies (sun, Earth, and moon).

Now, 'apogee' is not quite as exciting as perigee, as the Moon in apogee appears about the same size as usual. When an apogee occurs on a full moon, we call this a "micro" moon.

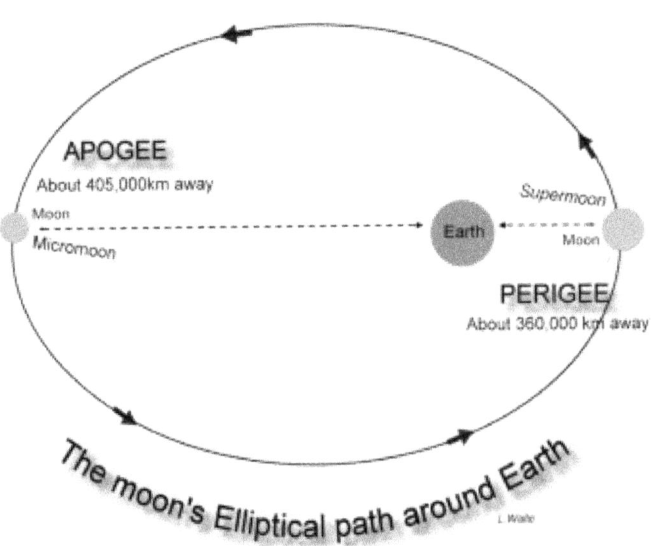

Perigee and apogee occur as part of the Moon's 27.3-day orbit around Earth, sometimes one or the other can happen twice in one month.

On the first of January 2024, we had an apogee with the Moon at 404,910 km from Earth.

On January 29, 2024, we experienced another, 405,780 km from Earth.

October 2024, will also experience two. In all other months, there will be only one. There will also be a perigee in each month of 2024, but June will have two, one on the 2nd, and another on the 27th.

Aphelion and Perihelion

From the Greek words 'Apo' meaning away, 'Peri' meaning near, and 'Helios' meaning 'sun.' Aphelion and Perihelion refer to the furthest and closest points that a planet, including Earth, orbits the Sun.

The rough average distance from the Sun to the Earth is 149,597,871 km.

In space, we use this distance as a unit of measurement called an 'Astronomical Unit' or 'AU.' Mainly for measuring the distances between a planet and the Sun.

The AU was established in 2012 by the International Astronomical Union as a standard measurement in astronomy.

The Earth rotates on its axis at a speed of 1,669.97 km per hour completing its elliptical orbit of the Sun in about 24 hours, 1 Earth day.

However, aphelion and perihelion mean a planet's closest and furthest points in its orbital path around the Sun. In contrast, apogee and perigee mean the furthest and the nearest points of a satellite, or celestial object, orbit from a planet.

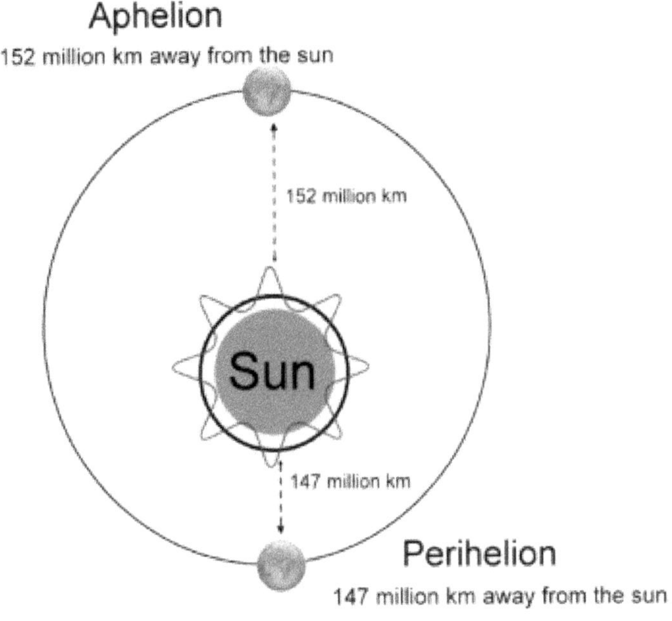

The closest we get to the Sun is 147 million km (perihelion). The furthest we get from the Sun is 152 million km (aphelion).

Equinox

I thought I would also touch on an Equinox, which has nothing to do with the Moon. An Equinox is when the Sun passes directly above the equator and occurs each year on or about March 21st and September 23rd. This is the only time the Northern and Southern hemispheres experience about the same amount of daylight.

Our equator is 0 degrees latitude during the equinoxes, and the sun's declination at noon is 0 degrees. This only occurs during the equinox and is the only time of the year when the Earth's axis is not tilting 23.5 degrees from the Sun.

An equinox occurs when the Sun's centre and the Earth's equator are aligned. Legend has it that during a spring equinox, you can stand an egg up on its end.

The ability to stand an egg up on its end during a spring equinox is a myth, or is it?

LUNAR ECLIPSES

Just a quick mention regarding lunar eclipses.

A few pages back we touched on this, but I thought that it needs just a little more explanation. So here it comes, when the Earth directly moves between the Sun and the Moon, it results in the Earth's shadow falling on the Moon, which is what we call a lunar eclipse.

A lunar eclipse can only happen during a full moon when the Sun, Earth, and Moon are all perfectly aligned.

The three types of lunar eclipses are a total eclipse, a partial eclipse, and a penumbral eclipse.

When the Earth's 'umbra' (the darkest inner part of Earth's shadow) covers the entire surface of the Moon that faces Earth during the eclipse, that is defined as a total eclipse.

During a total lunar eclipse, the Moon may take on a deep orange/red colour, due to the Earth's atmosphere bending the sunlight onto the Moon's surface, causing the unique view known as a blood moon.

Like a total eclipse, a partial eclipse is exactly that – only a section of the Moon passes through Earth's 'umbra,' leading to just a portion of the Moon being covered by the Earth's shadow.

Now, a penumbral eclipse occurs as the Moon moves through the Earth's 'penumbra' (a lighter outer shadow compared to the umbra).

Penumbral eclipses are typically less bright and not as impressive as the other types, making it somewhat challenging to differentiate from a regular full moon.

Lunar Eclipse - Old Wives' Tales

There are a few myths and wives' tales surrounding a lunar eclipse, including the following.

- *Pregnant women are to avoid a lunar eclipse or risk mental health issues to the unborn child.*
- *In some cultures, they avoid physical activity as they believe they will be injured.*
- *Avoid food and water or risk digestive problems.*
- *A lunar eclipse signals the wrath of the gods, a warning sign of disaster or war.*
- *In certain cultures, people practice meditation, prayer, or rituals during a lunar eclipse to ward off evil and negative energies.*

There are a few myths in religion as well, such as in Hindu mythology.

Rahu, one of the nine main celestial bodies and king of meteors, is said to eat the Sun, creating an eclipse.

In Vietnam, they believed solar eclipses were caused by a giant frog that eats the Sun. China had a dragon, and Norse people had wolves, all of whom allegedly tried to devour the Sun.

The Greeks thought the gods were angry and an eclipse was a sign of pending disaster.

Many cultures would make loud noises, such as yelling and banging pots together during eclipses believing that the noise would frighten evil spirits away. While some of the above practices may offer a mythical story or reason for an eclipse and even strategies to combat the demons through the production of loud noises, most are just folklore and wives' tales.

None are scientifically just. Likewise, no known medical conditions or risks are associated with a lunar eclipse.

A lunar eclipse is a predictable natural astronomical event, and with the correct eye protection, it has no negative impact on observers. An opportunity to witness one should be embraced and regarded as fortunate.

Rahu- is said to eat the Sun, thus creating a lunar eclipse.

THE BIRTH OF THE MOON

Composition, Formation and Theory

Of course, no Lunar eclipse could ever occur if the Moon did not exist. Yet it does and the mystery surrounding its creation has bewildered our greatest minds throughout Earth's history. However, over the years various theories trying to explain the formation of our moon have been put forward.

One such theory is the Capture Theory, which suggests that the Moon was a separate celestial body that was eventually captured by Earth's gravitational pull. According to this theory, the Moon formed elsewhere in the solar system and was later drawn into orbit around Earth.

Another theory, known as the Co-formation Theory, proposes that the Moon and Earth formed simultaneously from the same proto-planetary disk during the early stages of the solar system's evolution. This theory suggests that both bodies grew side by side, sharing similar building blocks and evolving together.

The Fission Theory suggests that the Moon was once part of Earth itself. According to this theory, Earth was spinning so rapidly (perhaps because there was no gravitational pull to slow it down yet?), during its early history that a large section broke off and eventually formed the Moon.

This idea is supported by the similarities in composition between Earth and lunar rocks.

Each of these theories has its strengths and weaknesses, and scientists continue to study them to gain a better understanding of how the Moon came into existence.

The Capture Theory, for example, fails to explain certain aspects of the Moon's composition that indicate a shared origin with Earth. On the other hand, the Co-formation Theory raises questions about why the Moon lacks significant amounts of volatiles compared to Earth.

Yet another theory (the one I like), is The Giant Impact Theory, which is one of the leading models explaining the formation of the Moon. According to this theory, a Mars-sized object collided with Earth during its early formation, causing a catastrophic impact.

This collision resulted in the ejection of a large amount of material into space, which eventually merged and amalgamated to form the Moon.

One of the key pieces of evidence supporting the Giant Impact Theory is the isotopic similarities between lunar and terrestrial rocks.

Isotopes are different forms of an element that have the same number of protons but differ in the number of neutrons. By comparing the isotopic composition of lunar rocks brought back by astronauts during the Apollo missions with those found on Earth, scientists have found striking similarities. This suggests that the Moon and Earth share a common origin. The collision that gave birth to the Moon had a profound impact on Earth's evolution.

The energy released during the impact caused both Earth and the colliding object to melt partially. As a result, the Moon formed from the material that was ejected and then combined. This process took place over millions of years.

By examining and evaluating these theories, scientists are gradually piecing together a more complete picture of the Moon's formation. Recent discoveries related to lunar formation have also further expanded our knowledge.

For example, did you know that recent evidence suggests that water may be more abundant on the Moon than previously thought? This certainly challenges the previous assumptions about its dryness.

Ongoing projects continue to investigate these findings and uncover new information about the birth of our closest celestial neighbour. As scientists advance in our understanding of lunar formation and composition, they continue to push boundaries in their exploration of this fascinating celestial body. By combining data from past missions with cutting-edge technology and analytical techniques, researchers are uncovering remarkable details about how the Moon may have come to be and its relationship with Earth.

So far, from the analyses of data and information gained through studying lunar samples collected and brought to Earth by the Apollo missions, scientists have learned a great deal and are starting to comprehend what makes the Moon tick, so to speak.

The Moon is primarily composed of rocks and minerals, like those found on Earth. It has a thin crust made up largely of silicate minerals, a mantle layer beneath it, and a small metallic core. The composition of these layers provides valuable clues about the Moon's geological history.

Lunar rocks, such as basalts and breccias tell us about volcanic activity on the Moon in its early history and provide evidence of impacts from meteorites.

The regolith, the layer of loose soil and debris covering the Moon's surface, is also crucial in understanding its composition. Analysis of regolith samples has revealed a mixture of fine dust particles and larger fragments, including tiny glass beads formed by ancient volcanic eruptions. The regolith also contains various elements and minerals, such as oxygen, silicon, aluminium, iron, titanium, and calcium.

Regardless of how it came to be, the formation of the Moon also had important consequences for Earth's development. The collision (that may have happened) caused Earth to tilt on its axis, creating the seasons we experience today. Additionally, it influenced the planet's rotation rate and shaped its geology.

The Moon's composition holds valuable clues about its history and its development. The more information we can collect, the more likely we will be able to determine the different layers that make up the Moon. It may also assist us in identifying what part the Earth played, if any.

THE MOON'S LAYERS

The Moon consists of several distinct layers, including the crust, mantle, and core. The crust is the outermost layer, composed primarily of rocks such as basalt and anorthosite. Basalt, a volcanic rock formed from solidified lava, is abundant on the Moon's surface and is responsible for its dark colour.

Anorthosite, on the other hand, is a light-colored rock that makes up the highlands of the Moon. Beneath the crust lies the Moon's mantle, a layer made up of denser rocks. The mantle extends downward to a depth of about 1,000 km (620 miles) and contains minerals such as pyroxene and olivine.

These minerals help provide some understanding of the temperature and pressure conditions within the Moon during its formation. At the Moon's centre lies its core, which is believed to be relatively small compared to Earth's core. The core is composed mainly of iron and nickel and contributes to the Moon's overall density. Studying the Moon's core can provide valuable information about its magnetic field and internal dynamics.

In addition to understanding the Moon's layers, researchers have also examined the presence of various elements and minerals on its surface. These include oxygen, silicon, magnesium, calcium, aluminium, titanium, and iron. These elements not only offer insights into the Moon's geological history but also serve as potential resources for future lunar exploration and colonization efforts.

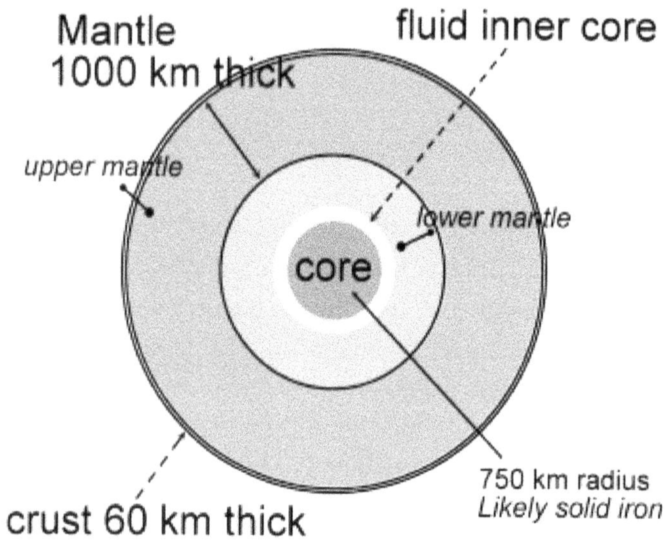

The data that is being collected, and the ongoing research are yet another example of important components in not only lunar exploration but in further understanding of the universe.

Each discovery provides further clues on the creation and constitution of our moon and Earth as well as other moons and planets within our solar system and beyond.

Now, the two common types of rocks found on the Moon include the previously mentioned basalts and breccias. Basalts are volcanic rocks that make up the majority of the Moon's surface. They are formed from hardened lava flows and have a dark, greyish colour.

Basalts are similar in composition to those found on Earth, suggesting they share a common origin.

Photograph by the Author

Breccias are rocks that are made up of fragments of other rocks fused. Their formation is through impact events where the intense heat and pressure cause rocks to shatter and then amalgamate.

Lunar breccias contain a mixture of different rock types, including basalt fragments and even fragments from Earth that were ejected by meteorite impacts.

Exploration of Regolith

As mentioned, regolith is the layer of loose soil and debris covering the Moon's surface and is made up of fine dust particles, small fragments of rock, and larger boulders. It is formed through a process called "gardening," where meteorite impacts break down rocks into smaller pieces over time.

Examining lunar rocks and regolith has uncovered yet more crucial information about the Moon's formation and its connection to Earth. By analysing lunar rocks and regolith samples brought back by various Apollo missions and robotic missions like Luna and Chang'e, scientists have been able to gain valuable insights into the Moon's formation and evolution.

The study of the isotopic composition of lunar rocks has helped confirm the Giant Impact Theory by showing similarities between isotopic signatures of certain elements in lunar and terrestrial rocks.

This suggests that the Moon was, shall I say, 'more than likely,' formed from material ejected during a massive impact between Earth and a Mars-sized object (the Giant Impact Theory).

Despite this evidence, it is not 100% certain this was how the Moon was created, it is still just a theory.

The composition of the regolith can vary depending on its location on the Moon. In some areas, basaltic material is the common type of material found, while in others it contains more exotic elements such as titanium and helium-3. Some regolith can also contain traces of water (ice) as found in the permanently shadowed regions near the Moon's poles. Additionally, further examination of lunar rocks has provided evidence of volcanic activity on the Moon in its early history.

The presence of basalts suggests that there were once active volcanoes erupting lava onto the lunar surface. Further analysis of lunar regolith has also revealed information about the effects of solar wind on the Moon's surface.

We know that regolith on the Moon contains microscopic glass beads known as agglutinates, which are formed when high-energy particles from the Sun strike lunar soil. These agglutinates, provide important clues about past solar activity and help scientists reconstruct the Moon's history of exposure to space weathering.

The examination of lunar rocks and regolith has uncovered some key elements that make up the Moon's surface and its past geological processes.

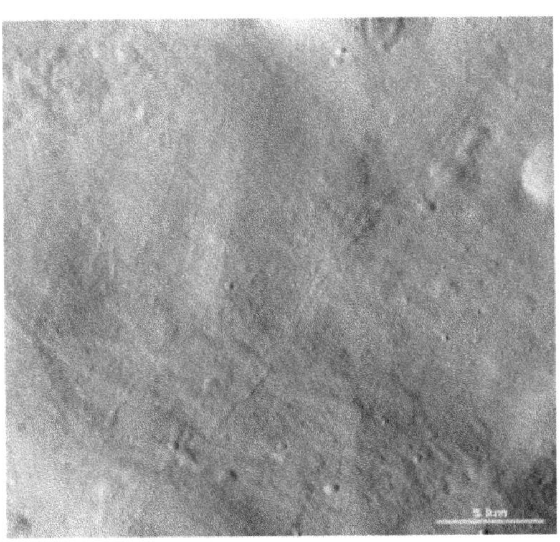

This image of asteroid Vesta from NASA Dawn spacecraft shows a particularly smooth part of the giant asteroid's surface, likely covered in fine-grained regolith material. - NASA

Scientists have made significant strides in identifying and understanding some of the mysteries surrounding the birth of our celestial neighbour.

One recent discovery relates to the Moon's water content. It was previously believed that the Moon was bone-dry, but studies conducted by lunar missions have found evidence of water molecules on its surface.

This finding has raised new questions about how water might have been delivered to the Moon and what impacts it may have on future exploration and potential colonisation.

Water, being a crucial resource for humans, is a vital element for any long-term visits to the Moon.

Another ground-breaking finding is the presence of volatiles, such as hydrogen and helium, within the Moon. These volatile elements were detected in samples collected during lunar missions, challenging previous assumptions about the Moon's composition.

Ongoing research programs are also shedding light on the Moon's geological characteristics. For example, scientists are studying moonquakes, which are seismic events occurring on the lunar surface.

By analysing data from seismometers left behind by Apollo missions and more recent lunar missions, researchers are gaining a better understanding of the Moon's internal structure and processes.

High-resolution imaging techniques have allowed scientists to closely examine the Moon's surface features and analyse its topography in unprecedented detail. This has led to the discovery of previously unknown geological formations and enhanced our understanding of lunar geology.

New analytical techniques and state-of-the-art instruments are being employed to study these samples, providing valuable insights into the chemical composition and age of lunar rocks.

Such research not only contributes to our understanding of the Moon's formation but also provides clues about Earth's early history.

From the detection of lunar water and volatiles to advancements in studying moonquakes and analysing lunar samples, each finding adds another piece to the puzzle of the Moon's origin. As technology continues to advance, we can expect even more exciting discoveries that will further illuminate the captivating story of how our closest celestial neighbour came to be.

Modern, inexpensive telescopes provide the capability for the average person to observe and study our moon.

The following photograph of the Moon was taken from my backyard observatory, near Brisbane, Australia.

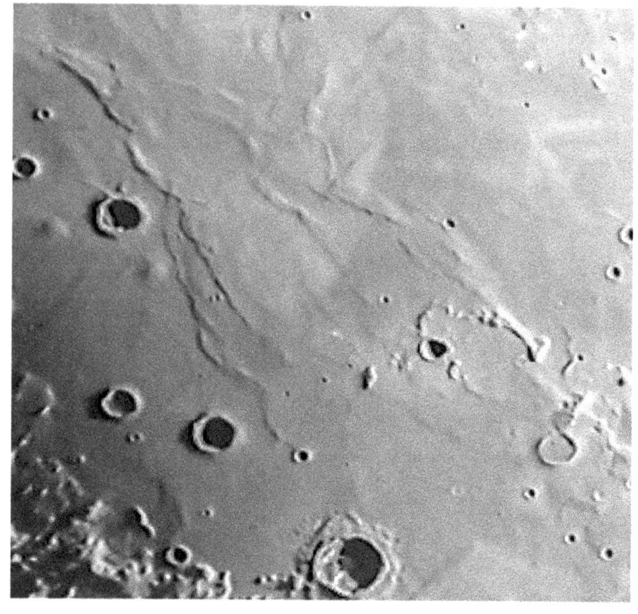

Photograph by the Author

Moon samples Apollo X1

This is the first lunar sample that was photographed in detail in the Lunar Receiving Laboratory (LRL) at the Manned Spacecraft Center (MSC).

"The photograph shows a granular, fine-grained, mafic (iron magnesium-rich) rock. At this early stage of the examination, this rock appears like several igneous rock types found on Earth. The scale is printed backwards due to the photographic configuration in the Vacuum Chamber." - *NASA*

MOON AND TIDES

The Gravitational Force of the Moon on Earth

As touched on earlier, the Moon's influence on Earth's tides is a result of the gravitational forces between these two celestial bodies. Gravity is a fundamental force that exists between all objects with mass, and it plays a crucial role in shaping the interaction between the Moon and Earth. When it comes to gravitational forces, the Moon is particularly significant due to its proximity to our planet. Its gravitational pull affects not only the Earth itself but also the waters that cover approximately 70% of its surface.

So how does this affect us?

To understand how gravity influences tidal formation, let's break down and examine the concept of tidal bulges a little more. There is an intricate connection between our delicate planet and the celestial objects that surround it. The gravitational pull from the Moon as it orbits Earth creates fluctuations in water displacement in large bodies of water. This in turn creates two tidal bulges, one facing the Moon and one on the opposing side.

Now, those locations on Earth that have the bulging effect, will experience higher water levels, whilst the other areas will experience lower water levels. The gravitational force is weakest at these locations due to its direction being opposite to that of the Moon's pull. These are our high tides and low tides.

This phenomenon occurs because Earth and its oceans are not perfectly rigid, allowing them to respond to gravitational forces. This interplay between gravity and tidal bulges gives rise to what is known as tidal forces.

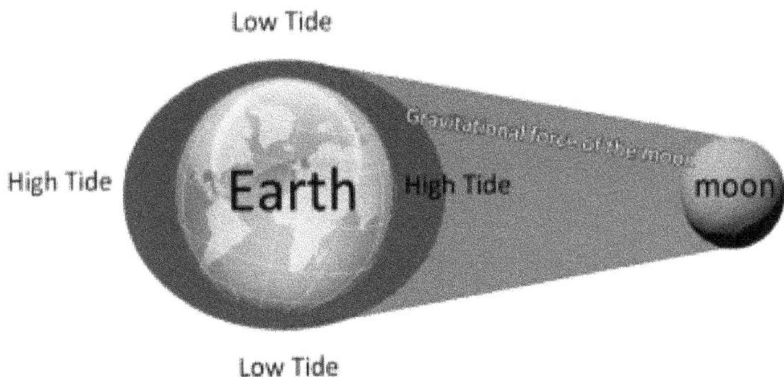

Tidal forces refer to the differential gravitational forces experienced by different parts of an object subjected to a gravitational field.

The interaction between tidal forces and Earth's rotation further influences tidal patterns. As Earth rotates within its 24-hour cycle, different locations experience changes in their alignment with the Moon, resulting in multiple high and low tide cycles within a day.

These cycles can vary in intensity depending on factors such as geographical features, weather conditions, and other astronomical effects.

Rocky Shoreline at low tide -Photograph by the Author

Gravity's role is necessary for the formation of tides and their influence on coastal areas, marine ecosystems, and even climate dynamics. It enables us to monitor and predict phenomena such as storm surges, coastal erosion, and even nutrient transport along coastlines.

Beyond Earth, gravity-driven tidal events can also be observed on other celestial bodies with moons orbiting them. For example, Jupiter's moon 'Io,' experiences intense volcanic activity caused by tidal forces generated by Jupiter's massive gravitational pull. Jupiter's gravity also has some effect on Earth, protecting us from comets for instance, but that's a rabbit hole we won't be going down in this book.

Exploring these extra-terrestrial tides provides valuable clues into the geological and dynamic processes occurring in these worlds. By understanding the intricate gravitational tango between the Moon and Earth, we gain a deeper understanding of tides' origins and behaviour.

This knowledge not only increases our scientific understanding but also has practical applications in numerous fields. From coastal management to space exploration, the influence of gravity on tidal formation carries significant implications for our planet and beyond.

Note: By the way, did you know that Jupiter has 95 moons? If you think that's a lot, Saturn has about 146 orbiting moons!

According to NASA, the number of moons in our solar system is about 293. I say that because this number seems to be continually increasing as we discover more.

Planets and their satellites

- Mercury 0 moons
- Venus 0 moons
- Earth 1 moon
- Mars 2 moons
- Jupiter 95 moons
- Saturn 146 moons
- Uranus 28 moons
- Neptune 16 moons
- Pluto (dwarf) 5 moons

Saturn is the winner with over 140 moons!

The photograph of Saturn was taken from my backyard with a Sky-Watcher 180mm Maksutov telescope on an equatorial mount.

Each year brings discoveries, such as more moons, contributing to our ever-expanding understanding of the celestial bodies within our cosmic reach.

Spring Tides

A Spring tide, also known as a 'King Tide', is aptly named due to its 'springing forth or rising of.' This tide usually occurs twice a month, typically during the full moon and new moon phases.

This is not a seasonal event, having no connection to the season of spring. Occurring twice each lunar month all year long. Seven days after a spring tide the gravitational forces from the Moon and the Sun align. This alignment results in higher high tides and lower low tides.

The characteristics of spring tides can vary depending on factors such as the distance between the Earth, moon, and Sun, and their relative positions as well as the local geographical landscape.

However, as a rule, spring tides have the greatest tidal range compared to other types of tides.

Neap Tide

On the other hand, another type of tide is a neap tide. Neap tides occur during the first and third quarter phases of the Moon (7 days after a spring tide) when its gravitational pull acts at a right angle to that of the Sun.

As the Moon's gravitational pull affects the water levels in the oceans, it creates a powerful force that leads to the rise and fall of tides. These tidal fluctuations have a significant impact on coastal environments. Coastal ecosystems, such as salt marshes, mangroves, and tidal flats, are highly influenced by the ebb and flow of tides.

During high tide, these areas become inundated with seawater, providing vital nourishment and habitat for a wide range of marine organisms. The incoming tide brings nutrients, detritus, and microorganisms into these ecosystems, supporting the growth of algae, sea grass, and more.

Tidal cycles also contribute to the transportation and distribution of nutrients along coastlines. As the tide recedes during low tide, it carries away excess nutrients and organic matter from coastal areas.

Understanding the factors behind the varying heights of spring and neap tides helps us comprehend the complex interplay between celestial bodies and our planet's oceans. By studying these tidal events, scientists gain valuable insights into Earth's dynamic systems and how they impact coastal ecosystems and human activities reliant on tidal patterns.

This process helps to maintain a balanced nutrient supply within these ecosystems, ensuring the availability of essential resources for marine life. Likewise, tidal currents influenced by tides create dynamic habitats that support a diverse array of species.

These currents can shape the physical structure of coastal areas, such as sandbars and estuaries, providing sheltered spaces and feeding grounds for many marine species.

Tidal flows also aid in the mixing of saltwater and freshwater in estuaries, creating unique brackish environments catering for some species that have adapted to fluctuating salinity levels.

Coastal processes driven by tides also contribute to shoreline erosion and sediment deposition. As tides push against the coastline, they can erode landforms over time, contributing to changes in coastal topography. Conversely, during periods of low tide, sediment carried by the tide may be deposited along the shore, leading to the formation of beaches and sandbars. The complex interplay between tides and coastal ecosystems supports a rich biodiversity and provides essential services for both marine and human life.

Coastal habitats influenced by tides serve as important breeding grounds for many commercially valuable fish species and provide critical feeding areas for migratory birds.

Additionally, these ecosystems contribute to carbon sequestration, shoreline protection from erosion, and buffering against storm surges.

By recognizing the intricate ecological relationships connected to tidal cycles, we can develop strategies to protect fragile coastal ecosystems and mitigate the potential impacts of human activities on these vital habitats.

Tides Beyond Earth

Tidal spectacles are not limited to just our Earth. Many other planets and moons also experience tides. Just as our moon experiences gravitational forces, moons orbiting other planets, are subject to the gravitational forces created by their parent planets. Tidal forces, and indeed gravity, help shape the surfaces and create the environments of these moons in spectacular ways. One example of extra-terrestrial tides can be seen on Jupiter's moon, Io.

This volcanic moon experiences intense tidal heating due to its elliptical orbit around Jupiter. Jupiter's gravitational pull, combined with the gravitational pull of its nearby moons, generates enormous heat within its moon Io. This causes Io's constant volcanic activity resulting in regular molten lava on its surface.

Another interesting example is one of Saturn's moons, Enceladus. Saturn's gravity and tidal forces inflict horrific stress on tiny Enceladus, causing the Moon to stretch and compress over and over.

This triggers the many powerful geysers on Enceladus to erupt violently. The water vapor and ice particles spewed up from the geysers are thrown into space contributing to the famous rings of Saturn.

We know this from studying some of the hundreds of moons in our solar system. Many of these have provided key information about their existence, such as surface elements, and their complex interactions with their parent planet contributing to their evolution.

LUNAR PHASES

The Festival of Light and Shadows

We have already lightly touched on this fascinating wonder that has puzzled humans for centuries, which is of course the lunar, or moon phases. I thought it appropriate to provide a little more information on this wonder that has been gifted to us by our neighbour.

So, in this section, we will explore the different lunar phases again, go over their dynamics, and discuss their cultural and scientific significance a little more.

Lunar phases refer to the different appearances of the Moon as it orbits around the Earth. The Moon goes through a continuous cycle of appearing to change shapes, ranging from a thin crescent to a full circle.

There are eight lunar phases, one coming after the other as it moves through its monthly cycle.

It takes the Moon 27.3 days to complete its orbit or Earth, however, due to the sunlight hitting the Moon, it takes 29.5 days to go from one new moon to the next, which gives us a month, thereabouts.

1. a new moon
2. a waxing moon
3. a first quarter
4. a waxing gibbous
5. a full moon
6. a waning gibbous
7. a third quarter
8. a waning crescent

The cycle of lunar phases begins with the new moon. During this phase, the Moon appears completely dark when observed from Earth. As the Moon moves along its orbit, a small sliver of light becomes visible, marking the start of the crescent phase. The crescent grows wider each night until it reaches the first quarter phase, where half of a new moon is illuminated. From there, the illuminated portion continues to expand, creating what is called the "gibbous" phase.

When the Moon reaches its final full moon phase, it appears as a complete circle when viewed from Earth. Then, the process reverses, and the illuminated portion gradually decreases until another new moon occurs. The three things behind the process of lunar phases are the positions of the Sun, Moon, and Earth with each other.

As the Earth orbits around the Sun, different portions of the Moon are illuminated by sunlight at different times. The Moon itself doesn't produce any light; it simply reflects the sunlight that falls upon it allowing us on Earth to see it.

For example, during a new moon, the portion of the Moon facing us is not illuminated by sunlight, making it appear completely dark.

This occurs because the side of the Moon facing the Sun is not visible from Earth. As the days go by, we start to see a thin crescent of light as the Moon progresses in its orbit.

This is known as the crescent phase.

Next comes the first quarter phase, where half of the Moon's face is illuminated by sunlight from our perspective. The term "first quarter" refers to the fact that we have completed one-fourth of a full lunar cycle. As we move further along in the lunar cycle, we enter the 'gibbous' phase.

The word "gibbous" means hump-shaped, and during this phase, more than half (but less than a fully illuminated portion) of the Moon is visible.

Finally, we reach the full moon phase when the entire side of the Moon facing Earth is illuminated by sunlight. This occurs when the Moon is positioned directly opposite to the Sun on Earth.

Whether you're observing a waxing crescent or a waning gibbous, understanding how these phases come about adds depth to our knowledge of our celestial neighbour and enhances our appreciation for its beauty.

External influences can have a significant impact on lunar phases, creating unique phenomena and altering their appearance.

Visual aids and diagrams are particularly helpful in understanding these concepts. They can illustrate how the changing angles and shadows create each phase, providing a clearer picture.

As the Moon orbits around the Earth, we observe different amounts of the illuminated portion from our vantage point. This creates what we commonly refer to as the phases of the Moon (see the above diagram).

One of the most notable influences is an eclipse, which occurs when the Earth, moon, and sun align in a specific way.

During a lunar eclipse, Earth's shadow obscures the Moon, causing it to darken or take on a reddish hue. This occurrence adds an enchanting element to the lunar phase, captivating observers around the world.

Atmospheric conditions can also affect the appearance of lunar phases. The thickness and composition of Earth's atmosphere can cause the

Moon to appear in different colours or even disappear completely during certain phases. For example, atmospheric particles can scatter shorter wavelengths of light, making the Moon appear redder during a crescent or gibbous phase. This phenomenon is often seen during sunrise or sunset when the Moon appears low on the horizon.

In addition to these influences, rare occurrences such as blood moons and blue moons add to the allure of lunar phases.

A blood moon refers to a total lunar eclipse where the Moon takes on a deep red or orange colour due to sunlight being refracted by Earth's atmosphere. These events are relatively infrequent and create a breathtaking spectacle for observers lucky enough to witness them.

In comparison, a blue moon is not blue but rather refers to an extra full moon that occurs within a calendar month. This phenomenon arises because the lunar month, which is approximately 29.5 days long, does not align perfectly with our traditional calendar month.

Though not as uncommon as a blood moon, a blue moon still holds cultural significance and is often associated with mystery and folklore.

Throughout history, cultural beliefs and superstitions have been intertwined with these rare occurrences. Various cultures have attributed special meanings or powers to blood moons and blue moons. Some believe they predict great changes or serve as omens for both positive and negative events. These beliefs have persisted over time, adding a sense of wonder and reverence to the study of lunar phases.

Lunar phases have held significant importance for various cultures, guiding their agricultural practices, aiding in navigation, and playing a pivotal role in ritual ceremonies.

Myths and legends were woven around these lunar phases, attributing divine meaning to each phase. For many, the waxing and waning of the Moon represented a cyclical pattern that mirrored the cycles of life, death, and rebirth.

In rural societies, lunar phases play a crucial role in determining optimal planting and harvesting times. Farmers observed the Moon to guide their agricultural practices, believing that specific lunar phases influenced plant growth and fertility.

For example, during a waxing moon, it was believed that plants would thrive and produce bountiful harvests.

Lunar phases also featured prominently in navigation. Sailors utilised the Moon's position in the night sky to determine their latitude and longitude, aiding them in safely navigating treacherous waters. By observing the changing shape of the Moon, sailors could accurately calculate their position at sea.

Rituals and ceremonies often revolved around lunar phases as well. Many ancient civilizations conducted rituals during specific lunar phases, believing they held spiritual significance. These rituals ranged from offerings to worshipping deities associated with the Moon, to elaborate ceremonies performed during full moons.

Even in modern society, lunar phases continue to hold cultural significance. For instance, some individuals associate a full moon with increased energy or changes in mood. In addition, various cultural festivals and celebrations still take place during specific lunar phases, preserving age-old traditions that honour the Moon's ever-changing appearance.

Many people also believe that certain lunar phases, although there is no scientific proof, can influence human behaviour. As we explore the wonders of lunar phases throughout history and across cultures, we gain a deeper appreciation for the profound impact they have had on human societies.

From agricultural practices to navigation techniques to spiritual beliefs, the influence of lunar phases has shaped our collective understanding of the celestial realm.

Often with a telescope, I prefer to look at the various phases of the Moon to see the detail of the terrain. A full moon is much too bright to permit good viewing. A pair of binoculars can also be used to see the Moon's terrain, particularly on the terminator line dividing day and night on the Moon.

Scientists study these lunar phases to gain insights into celestial bodies beyond the Moon. One method involves using lunar phases to measure distances in space.

By observing how celestial bodies appear from different positions on Earth during various lunar phases, astronomers can calculate their distances using the concept of parallax. Scientists analyse how lunar phases affect meteoroid impacts on the Moon's surface and study how variations in solar radiation during different phases impact space weather conditions.

By observing how the brightness of the Moon varies during different phases, scientists can gather data on the amount of sunlight reflected off its surface.

This information helps in monitoring solar activity, such as solar flares and coronal mass ejections, which can have significant impacts on Earth's magnetic field and climate. Additionally, lunar phases contribute to our understanding of meteoroids and meteor showers. During certain lunar phases, such as when the Moon is in its gibbous or full phase, its brightness can interfere with meteors and other astronomical observations. However, during new moon or crescent phases, when the Moon is less illuminated, astronomers have clearer skies and better opportunities to observe meteors streaking across the night sky.

Ongoing research continues exploring the scientific significance of lunar phases in various fields of study. Scientists are constantly developing innovative techniques and technologies to further our knowledge of celestial bodies by harnessing the power of lunar phases. These studies contribute to our overall understanding of space and help us unravel the mysteries of our vast universe.

Through careful observations and calculations during different lunar phases, astronomers can determine distances between objects, study lunar topography, monitor space weather, and investigate phenomena like meteoroids.

Lunar phases hold immense scientific significance in studying not only the Moon itself but also other celestial bodies.

FOLKLORE, STORIES, AND SONGS

Throughout various cultures of our history, the Moon has often been an essential character in traditional stories and myths that reflect man's fascination with heavenly bodies. These myths offer insight into how our ancestors viewed and revered the Moon. As a child, you may have seen the rabbit in the Moon's features or the 'man in the Moon' or listened to a nursery rhyme of a cow jumping over the Moon and of course, many of us have grown up with Jack and Jill, who went up the hill to fetch a pail of water. How is Jack and Jill relevant to the Moon you may ask? Well, there is a much older Norse tale that tells of Hyuki (Jack) and Bil (Jill) who went up, perhaps the same hill.

The original story (*short version*) went something like this.

Jack and Jill were sent up the hill to fetch a pail of water from a magical well. Water from this well was said to inspire poetry and prophecy. The man who carries the Moon across the sky each night in his chariot spotted the children. He scooped them up into his chariot and carried them away. The two children, with their buckets and poles, can still be seen there today.

The story of Hyuki and Bil is said to have originated in the 13th century by Snorri Sturiuson. It's uncertain if he was trying to personify the craters in the Moon or the Moon phases.

Another traditional story passed down through the generations is of Bahloo (the Moon), a native Australian legend that occurred back in the Dreamtime.

Bahloo, dropped down to visit the Earth and observed two young women near a river, they invited Bahloo into their canoe. Bahloo accepted but found it difficult to balance and fell into the river. Embarrassed by this, he hid.

He now shines brightly every month, but Bahloo remembers this little accident and shrinks away to gather his courage (moon phases).

The Moon was also of significant importance to the Native American Indians. They would tell time from one new moon to the next (lunar cycle). They would also name each month's moon to keep track of the seasons.

The Moon cycle for January was named the "Wolf Moon." This was because, on the cold nights of January, the wolves would howl hungrily outside Indian villages. February was the Snow Moon, and as you probably guessed, this was due to the heavy snowfalls.

Another name for one of the lunar cycles was the Crow Moon for March as the crows would caw as if to say goodbye to the cold weather.

April was the Egg Moon, May was the Flower Moon, June the Strawberry Moon, July was the Thunder Moon or Buck Moon (due to storms and buck deer would start forming antlers), August was the Red Moon, September the Harvest Moon, October the Hunters Moon, November the Beaver Moon, and December was known as the Cold Moon.

You can see that keeping track of the Moon's phases played a valuable role in the lives of the native American Indians. Many other cultures would reference or mark the Moon's phases as an important part of their daily lives.

Often, the Moon is featured as a divine presence or a celestial god possessing various mystical qualities and powers. It has also been commonly associated with fertility and cycles of life. In many other myths, the Moon is portrayed as a source of light to guide night travellers and sailors as well as visual comfort to lost souls.

These charming and sometimes obscure tales not only captivate our imagination but also provide insights into both the social and spiritual beliefs of our older cultures. Cultures that so many of us are now so far detached.

They tell us of the deep connections between humans and the Moon, emphasising its deep influence on human consciousness throughout our history.

Most myths are collections of stories passed down through generations and are about our understanding of our place in the universe and how we need to 'fit in' to survive.

Religious beliefs associated with the Moon have also existed throughout history, showcasing the significant role it has played in our spiritual requirements.

In different spiritual traditions, the Moon symbolizes different aspects such as femininity, fertility, and transformation. For example, in Greek mythology, Artemis was the goddess of the Moon and the hunt. She represented female empowerment and protection.

Some see the full moon as holding special significance and celebrate a full moon with rituals and prayers. Some cultures even perform ceremonial rituals during a lunar eclipse, viewing them as powerful events that can cleanse and transform believers.

These religious and spiritual beliefs that embrace the Moon serve as a reminder of the deep connection we have. Whether it is divine, a guide to survival in our daily lives, a source of comfort and inspiration, or a simple reminder of the cycle of life. Overall, the Moon is seen by most as a symbol of peace and enlightenment.

By contrast, exploring religious and spiritual beliefs associated with the Moon provides a rich understanding of how different cultures have perceived this celestial body throughout history. These myths, their meaning, and their reasoning can provide us with a much deeper appreciation for the Moon, and the effect it has had on us in shaping human culture.

But the Moon has also been a symbol of mystery and beauty, captivating the imaginations of artists, writers, and movie directors across different periods. In art, literature, and pop culture, the Moon is often represented in various ways, to portray unique meanings and interpretations.

The Moon often appears to illuminate the night sky in a variety of ways in both the visual arts and entertainment industries. Artists use various methods to express the glow of the Moon as either a surreal enchanting light or a sinister beacon of evil.

With well-chosen words, a series of brush strokes or clever photography techniques, an artist can convey an interpretation that triggers our imagination and fulfils the artist, author, or director's intention.

Of course, some extremely well-known works of art feature the Moon, such as "Starry Night" by Vincent van Gogh, and " Escape to Egypt " by Adam Elsheimer, both showing the Moon as a key subject of their paintings.

In 1711, artist ‘Donato Creti’, was commissioned by the Bolognese count, Luigi Marsili, to paint a series of small canvases titled “Astronomical Observations” which comprised 8 paintings depicting a variety of celestial objects. These were intended to be gifted to Pope

Clement X1. One such painting 'Moon', depicts two persons at night, one with a telescope pointed at a full moon. It is quite captivating.

Many people throughout history, including writers, poets, artists, and musicians, have been inspired and influenced by the fascinating glow and aura of the Moon, from its reflection in the water to its illuminating presence, the Moon has the power to awaken something within all of us.

The Moon is used as a medium of expression, symbolism or as a 'mood setter' in literature, film, and song.

Many artists refer to or use the Moon symbolically such as in William Shakespeare's poem "A Midsummer Night's Dream" with the Moon representing the passing of time and change.

Sylvia Plath's "The Moon and the Yew Tree" tells of a country graveyard at night, where the poet experiences an emptiness with the Moon. In William Wordsworth's, "Strange Fits of Passion I Have Known", Wordsworth was obsessed with the Moon in his poems, sonnets, and odes.

The Moon also has a universal allure, symbolizing beauty and wonder that transcends cultural beliefs and filters into all aspects of human expression, of course, this also means in song. There is an uncountable number of songs that refer to the Moon with their lyrics expressing desire, longing, or a sense of connection.

Some such songs, and there are many, include "Fly Me to the Moon", "Moon River", "Bad Moon Rising", "Moonlight Serenade," and even songs on Pink Floyd's "Dark Side of the Moon" album. These songs and music use the Moon as a powerful symbol to convey emotions and evoke a sense of soul searching.

The imagery accredited to the Moon in different contexts is complex yet simplified through the assortment of arts.

In the world of film, just like in music, the Moon often takes a prominent role, becoming a central element in the plot or a mesmerizing visual spectacle.

Whether it's in a sci-fi masterpiece like "2001: A Space Odyssey," where the Moon becomes a gateway to extra-terrestrial discovery, or in a romantic classic chick flick like "Moonstruck" which symbolises unexpected love, the Moon adds layers of meaning and symbolism to cinematic narratives.

Scientific advancements in space exploration and its influence have reshaped how the Moon is depicted in popular culture. As our understanding of lunar science expands, so do the creative possibilities for portraying the Moon in the media. In modern film, literature, music, and beyond, the Moon remains an enduring companion that continues to inspire artists.

It serves as a muse across diverse creative fields, exploring themes of love, mystery, and the spirit of exploration. As scientific knowledge evolves, so too does our appreciation of the Moon's significance, leading to fresh and imaginative portrayals in popular culture.

The Moon can represent femininity, intuition, and cyclical nature in many cultures. In mythology, figures such as Artemis or Chang'e are often associated with the Moon, further supporting its symbolic significance.

Additionally, the Moon is linked with renovation and rejuvenation, as it waxes and wanes in regular cycles reinventing itself with each new orbit.

While these symbolic representations may vary across cultures and periods, they all reflect humanity's fascination with the Moon's enigmatic beauty and profound influence on our collective imagination. Exploring this rich combination of moon symbolism, allows us to appreciate how this celestial body continues to inspire creativity and emotion in diverse artistic expressions throughout history.

Our world experiences and celebrates an assortment of events each year with numerous cultures holding festivals, rituals, and gatherings in honour of our moon. Many of these events take place each year aligning themselves with one of the Moon's phases such as a full moon, blood moon, or a lunar eclipse. By coming together to honour and appreciate the Moon, these celebrations serve to preserve cultural heritage and foster community bonds.

One notable example is the "Mid-Autumn Festival," lovingly known as the Moon Festival, cherished across East Asia, particularly in China, Taiwan, and Vietnam. This cherished event unfolds on or about the 15th day of the eighth lunar month, a time when the Moon shines brightest and fullest.

Families come together for a heart-warming reunion dinner, relishing the sweetness of mooncakes, a historic traditional pastry denoting harmony and unity. Lanterns adorn the night, their glow casting a magical aura during joyous processions.

In East Asian culture, this festival embodies values of togetherness, appreciation, and harmony. It's a special occasion for families to celebrate abundance and express gratitude for their blessings. Beyond its lunar associations, the Mid-Autumn Festival fosters stronger family bonds and fosters goodwill within communities.

Another popular, and traditional lunar festivity is the Dragon Boat Festival, observed in various East Asian nations like China, Hong Kong, Singapore, and Malaysia.

While this event primarily commemorates the ancient poet Qu Yuan, its roots intertwine with lunar reverence. Marked on the fifth day of the fifth lunar month, the festival features spirited dragon boat races, where teams row to the beat of thundering drums.

Beyond East Asia, diverse cultures celebrate unique lunar festivals. For instance, in India, Diwali dubbed the 'Festival of Lights', honours age-old rituals aligned with the lunar calendar. This grand festival sees millions of oil lamps illuminating homes, symbolizing the victory of light over darkness and good over evil.

During the Dragon Boat Festival, people indulge in zongzi, delectable sticky rice dumplings wrapped in bamboo leaves. These treats carry symbolic significance, warding off malevolent spirits and diseases.

Like their counterparts, the Dragon Boat Festival unites communities in spirited revelry and reverence for cultural heritage.

In Islamic tradition, the sighting of a new crescent moon marks the beginning of each lunar month. These sightings are crucial for determining dates related to Islamic holidays such as Ramadan and Eid al-Fitr.

Muslims worldwide observe these sacred occasions by fasting from dawn until sunset during Ramadan and then gathering for feasts and prayers to mark the end of fasting during Eid al-Fitr. These examples illustrate how different cultures have developed their distinct rituals and customs surrounding lunar events.

Whether it's through lantern processions, dragon boat races, or lighting oil lamps, these celebrations bring people together in unity and their cultural heritage while passing down stories, rituals, and traditions from one generation to another.

Through all these demonstrated examples of which there are many more, these meaningful gatherings centred around the Moon, people not only connect with their past but also forge bonds that strengthen their sense of belonging and shared identity.

LUNAR EXPLORATION

Historical Lunar Missions

The exploration of the Moon has been a fascination for humans throughout history. We will look at some of the earliest lunar missions in this section, including the Soviet Luna program and NASA's Ranger and Surveyor missions. These missions were critical in laying the foundation for future explorations and deepening our understanding of Earth's celestial companion. The Soviet Luna program, initiated in the late 1950s, aimed to achieve a series of firsts in lunar exploration. The Luna 2 mission, launched in 1959, became the first human-made object to reach the Moon's surface, crashing into it successfully. This mission provided valuable data about the Moon's surface composition and helped pave the way for future 'softer' landings. NASA's Ranger program followed soon after, starting in 1961.

The primary objective of the Ranger missions was to impact the Moon's surface with high-resolution imaging cameras capturing close-up pictures until impact.

Although the early Ranger missions experienced technical difficulties, later missions achieved success and returned valuable images of the Moon's surface.

The Huntsville Times

Man Enters Space

'So Close, Yet So Far,' Sighs Cape

U. S. Had Hoped For Own Launch

Hobbs Admits 1944 Slaying

Soviet Officer Orbits Globe In 5-Ton Ship

Maximum Height Reached Reported As 188 Miles

'To Keep Up, U. S. A. Must Run Like Hell'

Praise Is Heaped On Major Gagarin

'Worker' Stands By Story

Reds Deny Spacemen Have Died

Reds Win Running Lead In Race To Control Space

Soviet cosmonaut, Yuri Gagarin, became the first man in space on April 12, 1962. - NASA

The two Soviet crewmen for the Apollo-Soyuz Test Project mission are photographed at the launch pad at the Baikonur Cosmodrome in Kazakhstan on the morning of the Soviet ASTP lift-off on July 15, 1975.

They are cosmonauts Aleksey A. Leonov (left), commander; and Valeriy N. Kubasov, flight engineer. Leonov is waving to well-wishers at the launch pad. The Soviet ASTP launch preceded the American ASTP Apollo lift-off by seven and one-half hours. The American and Soviet spacecraft were docked in Earth orbit for a total of about 47 hours on July 17-19, 1975. NASA ID: S75-32343

Building on the information learned from these early missions, NASA launched its "Surveyor" program in 1966. These unmanned spacecrafts were designed to soft-land on the Moon and gather detailed information about its surface characteristics. The Surveyor missions provided crucial data on the Moon's topography, soil properties, and thermal conductivity.

They also demonstrated that a safe landing could be achieved in preparation for manned missions. These historical lunar missions played a vital role in advancing our knowledge of the Moon. They laid the groundwork for future explorations and provided useful information that would later be used to tackle some of the challenges involved in putting man on the Moon.

The Atlas-Agena-4 boosted the Ranger IV spacecraft for the first U.S. lunar impact on April 23, 1962.- NASA

The information gathered through these early missions was of tremendous assistance to us and led to the discovery of key scientific findings that provided invaluable clues into its geological history and evolution.

The instruments carried by lunar missions allowed scientists to collect more detailed data about the Moon's composition, topography, and physical properties than ever before. For example, seismometers were used to measure moonquakes, providing insights into the Moon's internal structure and seismic activity.

Spectrometers analysed the reflected light from the Moon's surface, revealing information about its mineral composition. Cameras mounted on the spacecraft captured high-resolution images of the Moon, enabling detailed mapping and analysis of its features.

One of the significant findings from lunar missions was the discovery of water ice in the polar regions of the Moon.

Analysis of data collected by spacecraft, like NASA's Lunar Reconnaissance Orbiter, revealed the presence of these ice deposits in perpetually shadowed craters. This finding has important implications for future exploration and potential resource utilisation on the Moon.

Additionally, lunar missions brought back many samples of moon rocks, (over 20 kg), allowing scientists to determine their age through radiometric dating techniques.

The analysis of these samples indicated that the Moon is approximately 4.5 billion years old, providing crucial insights into its formation and early history. The composition of these rocks also provided evidence for the theories about the Moon's origin, such as the Giant Impact Theory discussed earlier.

These missions have allowed us to identify impact cratering as a primary geological process on the Moon, shaping its surface over billions of years. Through using specialised instruments and equipment, scientists were able to gather vital data and make significant discoveries.

From confirming the presence of water ice in polar regions to analysing moon rock samples to determine their age, these missions have contributed invaluable information that helps us piece together the fascinating story of Earth's moon.

Photograph by the Author

CRATERS

The above photograph depicts the Tycho crater along with Plato and Copernicus, all of which are very prominent lunar features.

Features such as the maria, the highlands, and the rilles are fascinating, but my favourite feature on the Moon must be the craters. In my opinion, the craters are one of the most striking landscapes on the Moon's surface. From my backyard, I can observe all of them through a telescope, with the craters standing out prominently.

Craters are formed when asteroids or comets collide with the Moon, creating impact craters of various sizes. The largest and most well-known lunar crater is the South Pole-Aitken Basin, which stretches over 2,500 km (1,550 miles) in diameter.

This colossal impact basin brings valuable perspectives into the Moon's early history and its bombardment by space debris. Maria, (the dark plains on the Moon's surface) were once presumed to be bodies of water. These flat, smooth-looking areas were created by ancient volcanic activity that flooded the region with molten lava.

Scientists have gained significant insight regarding the Moon's volcanic past through studying these maria, and its interaction with the solar system. In contrast to the maria, the lunar highlands are rugged and mountainous regions that cover a significant portion of the Moon's surface. These elevated areas contain numerous craters, mountain ranges, and impact ejecta.

They provide a glimpse into the early stages of the Moon's formation and its intense meteorite bombardment billions of years ago. The highlands also offer breathtaking views of Earth from the Moon's surface.

'Craters are one of the most striking features on the Moon's surface'.

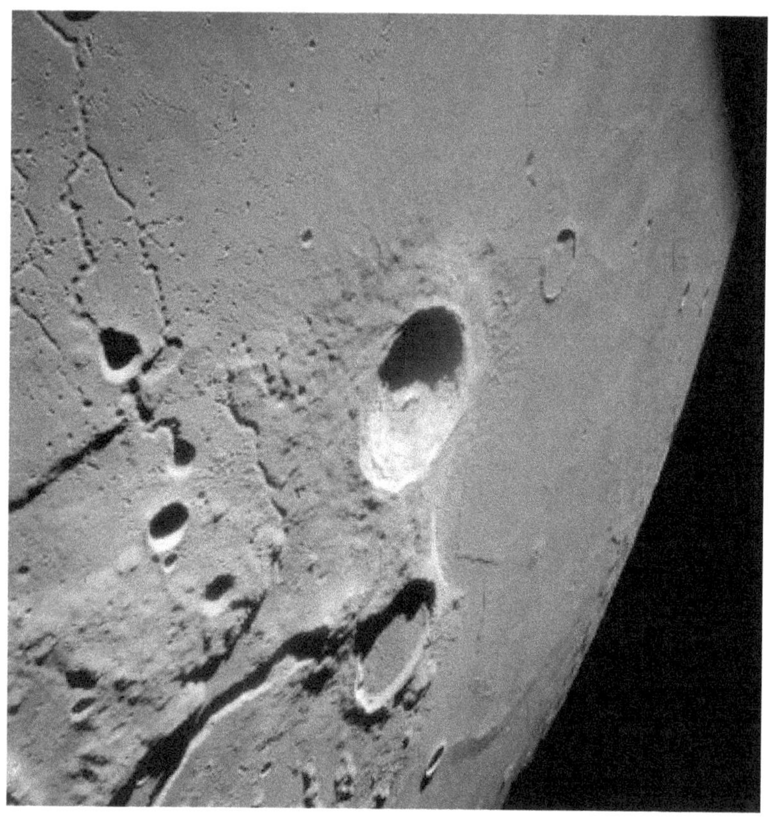

lunar orbit view: Craters Aristarchus & Herodotus -NASA

Rilles are long, narrow channels or grooves found on the lunar surface. These formations may result from ancient volcanic activity or tectonic processes. One notable example is Hadley Rille, located near the landing site of Apollo 15. This winding channel stretches for over 100 km (62 miles) and offers insights into the Moon's volcanic history and potential resources.

Certain lunar landmarks were specifically targeted for particular purposes. The Apollo missions, for instance, allowed astronauts to collect a variety of samples from different regions and study them back on Earth.

Notable Apollo landing sites include Tranquillity Base (Apollo 11), Descartes Highlands (Apollo 16), and Taurus-Littrow Valley (Apollo 17). These sites provided researchers with invaluable data about the Moon's composition and history and helped scientists piece together the Moon's geological puzzle. By analysing impact craters, volcanic plains, mountain ranges, and other formations, they can reconstruct its past and understand how it has evolved over billions of years.

These dark plains were once thought to be seas by early astronomers, hence their Latin name. These plains consist of contrasting highlands, and rugged terrains filled with craters and mountain ranges.

'Sea of Tranquility'- NASA/Unsplash

I think the acknowledgment of these landmarks is important. Through the samples and photographs collected, we established fundamental information that led us to our current understanding of the Moon's evolution.

As you read further, imagine yourself standing on the Moon's surface, surrounded by its awe-inspiring beauty. Picture the vast expanse of the lunar plains, known as maria, scattered across the Moon's surface.

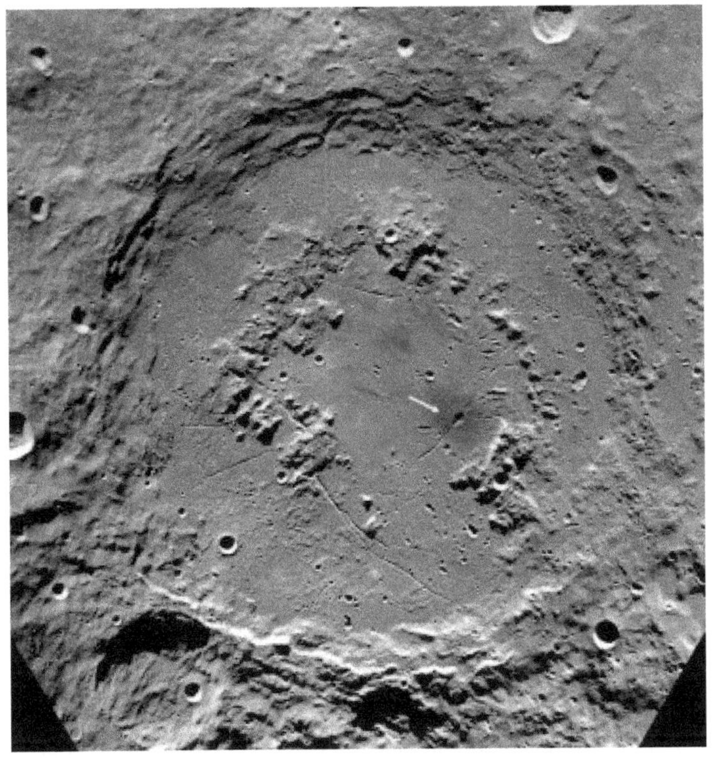

Schrodinger Basin -NASA

Schrodinger Basin (2) -NASA

The mind-blowing and intriguing geological formations have fascinated scientists for decades. From deep craters to winding channels (rilles), each feature tells a story about the Moon's geological processes and its turbulent past.

Tycho crater - NASA

On June 10, 2021, NASA's Lunar Reconnaissance Orbiter captured a dramatic sunrise view of Tycho crater. A very popular target with amateur astronomers, Tycho is located at 43.37°S, 348.68°E, and is about 51 miles (82 km) in diameter. The summit of the central peak is 1.24 miles (2 km) above the crater floor. – NASA

Quite a few lunar features and landmarks were photographed by the Apollo missions. From Tycho and Schrodinger, to Kings Crater and Goclenius.

KINGS CRATER -As16-122-19580

This vertical view, provided by NASA, shows the King Crater on the lunar surface. It was exposed with colour-positive film in a hand-held 70mm camera onboard the Command and Service Modules during the Apollo 16 mission's 98th orbit of the Moon.

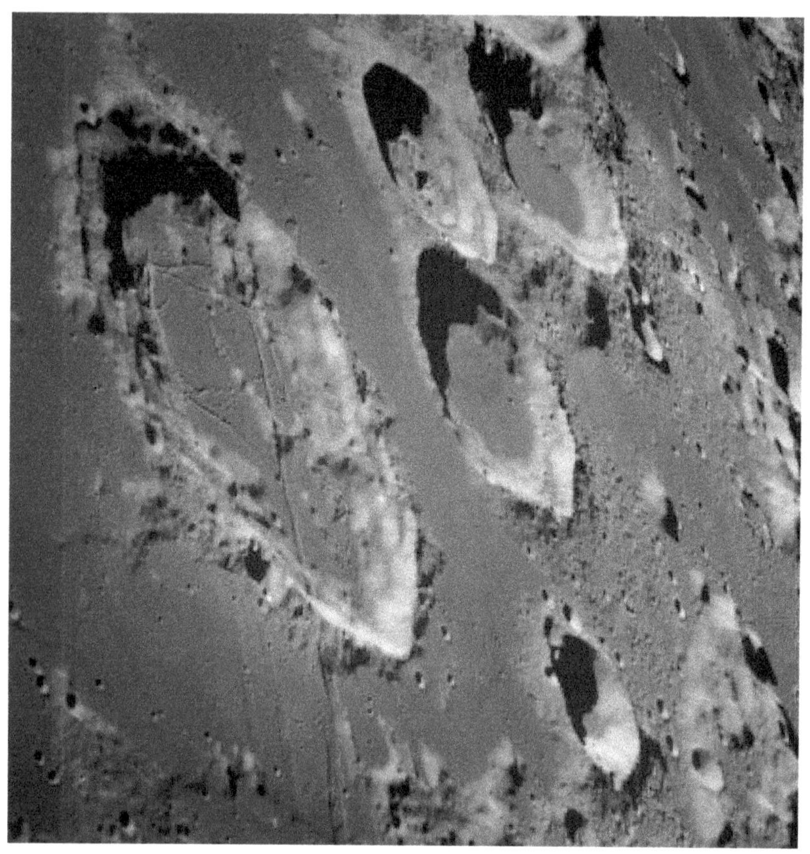

This photograph was taken from the Apollo 8 spacecraft with a long focal length lens, looking south at the large crater Goclenius, which is in the foreground. Hold a picture with Goclenius at the bottom centre. The three clustered craters are Magelhaens, Magelhaens A, and Colombo A. The crater at the upper right is Gutenberg D. Date Created:1968.- NASA

Schrodinger Basin, named after Erwin Schrodinger, best known for his mind-provoking thought experiment involving a cat, was a Nobel Prize-winning Austrian physicist.

Schrödinger's cat thought experiment illustrated the concept of quantum superposition and the interpretation of quantum mechanics. Schrödinger contributed to the fields of colour theory, thermodynamics, and molecular biology. In 1933 Erwin Schrodinger and Paul Dirac shared a Nobel Prize for their pioneering work in wave mechanics.

The Schrodinger Basin is a relatively new impact crater formed some 3.5 billion years ago and over 300 km in diameter. Its mountainous ringed peaks rise over 2 km from the basin floor. Due to its size and ring of peaks, it is considered an impact structure more than an actual crater. Schrodinger's Basin sits in a much larger and older known basin, the South Pole - Aitken Basin.

The South Pole Aitken Basin (SPA) - is located on the Moon's far side and is the largest, oldest, and deepest known impact crater. Stretching 2,500 km in diameter and 6 to 8 km deep the SPA was named after Robert Grant Aitken, an American mathematics and astronomy professor. In 1915 Robert Aitken discovered and measured the orbit of over 3000 double stars and was awarded the Bruce Medal for his work.

The South Pole Aitken basin is believed to be about 4.2 billion years old, making it one of the Moon's oldest craters. Initially discovered in the 1960s by one of the orbiting satellites, the full extent of the crater was not known until after the Clementine and Galileo missions in the 1990s. Recent evidence suggests an excess mass in the Moon's mantle under the South Pole Aitken basin, over 300 km deep. This stirs the science community in establishing how the impact crater was formed.

The mass anomaly suggests that the basin floor was not caused by melt sheet contractions as with other basin floors.

The South Pole Aitken Basin is a fascinating study and is currently considered the largest known structure in our solar system.

King (or Kings) Crater, is over 75 km in diameter and 5 km deep King Crater was formed through an asteroid or comet impact about 4 billion years ago. It is a "classic complex" crater, meaning that gravity would have caused the initial crater walls to collapse, forming a complex structure with a central peak and shallow diameter. In 1970 the crater was named from Crater 211 to King Crater after American physicist and astronomer Arthur Scott King (1876-1957) and American astronomer Edward Skinner King (1861-1931). Unfortunately for astronomers, it is also on the Moon's far side.

Tycho, this crater was named after Tycho Brahe (1546-1601), a Danish astronomer in the 16th century. Tycho is quite noticeable in the Moon's southern highlands and is an impact crater visible from Earth. It is also a popular target for lunar observers with a small telescope. The crater's 5 km high continuous wall encapsulates the 100-million-year-old crater. At approximately 80 km wide, Tycho has a mountain or 'peak' right in the middle. Tycho is a large bright crater, with a ray system having some sections reaching 1,500 km in length. When I look at my photographs, I often see the Moon resembling a rockmelon or cantaloupe, with Tycho Crater resembling its base.

Plato, named after the Greek philosopher, is an impact crater with a diameter of about 100 km. Its crater has been filled with about 2.5 km of lava that buries its peaks. Another great one to observe with binoculars or a small telescope.

Copernicus was named after Nicholas Copernicus (mathematician and astronomer), who demonstrated to the world that Earth revolved around the Sun. This crater sits about in the centre of the Moon's disc and is easily spotted.

Less than a billion years old, its diameter is about 90 km and light in colour as the solar winds have not yet had the chance to take their toll. Another couple of billion more years should darken it up!

FUTURE EXPLORATIONS

Our collective exploration efforts will not only reveal more about the Moon itself but also ignite our imagination for what lies beyond the cosmos.

As humanity continues to explore the wonders of space, our attention remains fixed on our closest celestial neighbour, the Moon.

The next exciting chapter of lunar exploration is about to unfold with planned upcoming missions that will further expand our understanding of this mesmerising landscape.

One significant mission on the horizon is NASA's Artemis program. Named after the Greek goddess associated with the Moon, Artemis aims to return humans to the lunar surface by 2024, nearly five decades after the Apollo missions. This ambitious endeavour seeks not only to establish a sustainable human presence on the Moon but also to pave the way for future missions to Mars and other destinations in our solar system.

The primary goal of Artemis is to land astronauts, including the first woman and the next man, near the lunar south pole. This region with its permanently shadowed craters, is believed to contain vast deposits of water ice, a critical resource for sustaining human life and potentially producing rocket propellant. By harnessing these resources, future explorers could establish a lunar outpost or even launch missions deeper into space.

In addition to NASA's efforts, other countries and private companies have set their sights on the Moon.

For example, China's Chang'e program, named after a mythological moon goddess, has already achieved several milestones, including robotic landings and the successful return of lunar samples. China plans to continue its exploration with future missions that include sample returns from unexplored regions and potential crewed landings. International collaboratives such as the Lunar Gateway project involve partnerships between NASA, the European Space Agency (ESA), the Canadian Space Agency (CSA), and others. The Lunar Gateway is envisaged as a small space station orbiting the Moon that will serve as a staging point for lunar surface missions.

This collaborative effort aims to facilitate global cooperation and knowledge-sharing to advance our understanding of the Moon. To support these future explorations scientists and engineers are working diligently to develop new technologies and capabilities. Advanced rovers and landers equipped with state-of-the-art scientific instruments will enable closer examination and analysis of the Moon's surface.

These instruments will be able to collect and analyse more samples from the Moon's surface than we have ever been able to collect before. This could provide more in-depth data on the Moon's geology, chemistry, and potential past or even present life signs.

With each new mission, we inch closer to solving more mysteries concealed within the Moon's landscape. The planned and upcoming lunar missions represent our continued dedication to exploring beyond Earth and expand upon our determination in space exploration. Our collective exploration efforts will reveal more about the Moon and ignite our imagination for what lies beyond the cosmos.

Together, we stand at the threshold of a new era of discovery, ready to unveil the secrets that have long been concealed beneath the Moon's peaceful glow.

Photograph by the Author

HUMANITY'S FIRST STEP

The Apollo Missions

One of man's most significant achievements in history has been the ability to set foot on another celestial body, our moon.

Early space exploration through NASA's Apollo program was instrumental in this achievement with Apollo 11 delivering humans to the Moon's surface.

The initial objective was more of a response to the then-Soviet Union's space missions having already sent a man, Yuri Alekseyevich Gagarin, into space. Gagarin completed one orbit of Earth in the Vostok 1 craft on April 12, 1961.

If not already in existence, the space race between the USSR and the US was certainly activated after a speech delivered by US President John F Kennedy, on September 12, 1962. The President gave a speech at Rice University on the nation's space effort, to gain public support for the Apollo program. In that speech, he committed to sending a man to the Moon before the decade's end. The rest is history.

The plan was to prove American dominance in technology to the world by establishing a human presence on the Moon. Consisting of several missions, the Apollo program launched its first unmanned flight in February 1966.

January 27, 1967, a tragic accident killed astronauts Virgil Grissom, Edward White and Roger Chaffee after a fire broke out on Apollo 1 during a practice training session.

Over the next two years, several more rehearsals and unmanned test flights were conducted until 1969 when NASA was ready to send a man to the Moon.

A little after 2 am July 20, 1969, Astronauts Commander Neil Armstrong, Michael Collins, and Lunar Module Pilot Buzz Aldrin, sitting on top of a Saturn V rocket, were launched into history. The three-day, three-hour, and 49-minute journey was by no means an easy feat and involved several precise manoeuvres including the 'lunar orbit insertion' manoeuvre, which would see them enter the Earth's orbit temporarily before placing them in a trajectory aligning them to the Moon.

Once in the Moon's orbit, the Command Module would orbit the Moon 31 times before making the journey back to Earth.

Commander Armstrong and Astronaut Collins, now inside the Lunar Module, separated from the Command Module, made their descent to the lunar surface.

Their mission was to complete the task set by President Kennedy in 1961 by landing on and returning from the Moon. They also had flight objectives and tasks such as setting up experiments and gathering samples.

Two astronauts walked about on the lunar surface after landing a little off course exploring and traversing the rugged terrain and examining the different geological features. This would have been the most exhilarating experience one could ever imagine.

They would have felt the benefits of the reduced gravity on their bodies as they walked (bounded) around setting up equipment, conducting tests, and noting and photographing their observations.

The Moon's gravity is approximately six times weaker than Earth's and comes with some advantages. You see, their specialised space suits weighed in at about 91 kg (200 lbs.) on Earth. Thanks to the Moon's gravity, this was reduced to about 13.5 kg (30 lbs.) when they were bounding along collecting their samples.

Not only that, but they would have set some world records when it came to the high jump with a staggering 3 metres (9.8 feet) high!

The Apollo missions have certainly played an integral role in not only moon exploration but space travel. Through ground-breaking research and innovation, scientists were able to not only land man on the Moon with Apollo 11, but also collect and analyse valuable data about the Moon's exosphere, geological constitution, and magnetic field.

They were also able to further understand the effects of the Moon's gravity on Earth, as well as hypothesising the uses of the Moon as a possible launching stage for future explorations.

In addition, the tremendous technology and ingenuity that was developed and utilised for the Apollo missions have had a lasting and intensifying effect on our thirst for knowledge, artificial intelligence, and faster, better technology for everyday tasks.

These developments were fundamental in a variety of many things including our recent and progressive digital transformation as well as leading to the production and development of microelectronics, better energy storage, and of course some impressive advanced rocketry and spacecraft. We also have continued to improve upon solar panels and their efficiency as well as medical devices and optics.

The impact of the Apollo program on humanity is enormous, yet many are unaware of its benefits.

Putting a man on the Moon demonstrated the capability of human ingenuity, cooperation, and determination. The event inspired the imagination of people around the world creating a sense of unity, pride, and accomplishment.

The memorable image of man stepping onto the lunar surface for the first time, and the photographs of astronauts planting a flag became powerful symbols of our achievement.

It is worth recognizing the success of the Apollo program and how it inspired generations of astronomers, engineers, and astronauts who continue to push the boundaries of space exploration. It also laid the foundation for subsequent missions and programs, such as the Space Shuttle program, and the International Space Station, and plans for manned missions to Mars.

The lessons we learned, and the knowledge we gained from the Apollo missions certainly paved the way for future ventures and ignited our ambition to explore the unknown frontiers of space.

The Apollo Spacecraft

The Saturn V rocket and the Apollo spacecraft were the dynamic duo that took man to the Moon and back.

Apollo 11 stood some 111 metres (363 feet) tall as it docked on the launch pad and consisted of a series of engines in three separate stages. Stage one had five, F-1 engines and was a powerhouse with an incredible 7.5 million pounds of thrust.

Stage two had five J-2 engines and Stage 3 had only one J-2 engine.

On top of this powerhouse, was the actual spacecraft, or capsule, which housed four integral compartments. The Command Module, the Service Module, the Lunar Module, and the engine.

The Command Module "Columbia," was the only part of the machine that would return to earth with the astronauts. A snug little cone-shaped cabin about 4 metres wide and 3.5 metres tall, was the home for three astronauts for a little over 8 days before splashing down in the Pacific Ocean some 1,528 km (920 miles), southwest of Honolulu on July 24, 1969, at 5:50 AM.

The President of the United States at the time, Richard Nixon, was on board the USS Hornet to welcome the returning astronauts. The three men, the Command Module Columbia, and the 22.2 kg (49 lbs.) of lunar samples were all kept in quarantine for 21 days. Although there are some reports that they were allowed out a few days earlier.

The Service Module was a 2.5cm thick aluminium (honeycombed panels) cylinder with a diameter just short of 4 metres and about 7.5 metres long.

This module was in two parts, one being the propulsion unit, and the other housed critical subsystems, fresh water supplies and consumables.

By the way, the Service module burned up upon re-entry.

The 'Lunar Module'- "Eagle" was at the rear of the capsule and was a two-stage vehicle designed specifically to transport two men to the Moon's surface and for lunar explorations near and on the Moon.

The Lunar Module weighs approximately 15,100 kg (33,000 lbs.) and contains several self-contained experiments that would be deployed and left on the Moon's surface once initiated.

It carried the two astronauts, Armstrong, and Aldrin, to the lunar surface whilst Collins stayed with the Command Module orbiting the Moon about 30 times at an altitude of around 100 km (75 miles).

After 21.45 hours, Armstrong and Aldrin were reunited with Collins back in Columbia (Command Module), and the Eagle was left to orbit the Moon.

Is it possible that it could still be in orbit today?

Apollo 11 moon landing occurred on July 21st, 1969. -NASA

The Apollo missions were truly an extraordinary feat of human ingenuity and determination, and it is worth noting that the significance of the Apollo missions cannot be overstated. We have derived so many benefits from the learnings and technology due to moon exploration.

The technologies developed for the Apollo missions had a far-reaching impact.

This photograph by NASA, Shows Astronaut and scientist Harrison Schmitt standing next to a boulder during the third EVA on the 13th of December 1972.- Apollo 17

LUNAR SCIENCE

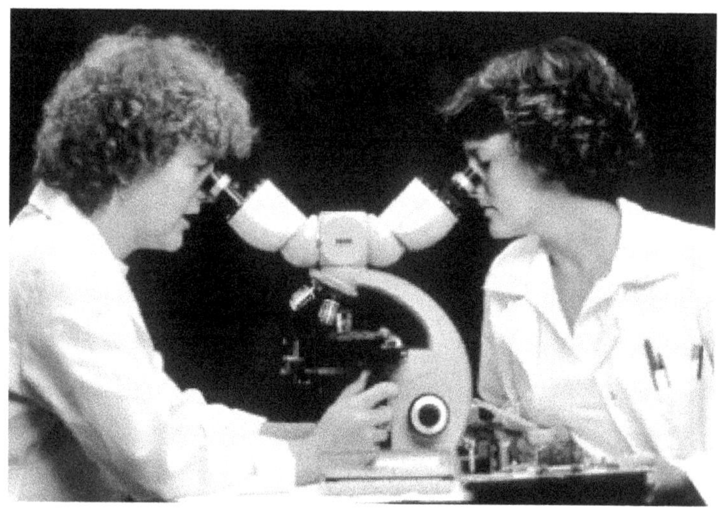

Since humanity first set foot on the Moon, studying it has been a cornerstone of our exploration efforts, helping us unlock its secrets and deepen our grasp of the wider solar system. This multidisciplinary field, spanning geology, chemistry, physics, and astronomy, continues to push boundaries.

With meticulous scientific approaches and cutting-edge technology, researchers can peer into the Moon's makeup, history, and atmosphere. They employ diverse methods and tools, like remote sensing through telescopes and spacecraft, to gather data thousands of lightyears away. Through this, they discover surface features, map topography, and identify geological formations.

Additionally, spacecraft missions have also been pivotal, fetching physical samples from the Moon's surface for Earth-based scrutiny. This not only sheds light on the Moon's distinct traits but also enriches our comprehension of the solar system. By dissecting its formation and composition, scientists garner insights into the mechanisms shaping our planet.

I think of the Moon as a sort of time capsule, recording the early solar system's saga and offering us glimpses into billions of years of evolution.

By analysing the impact size, and geomorphology of impact craters on the Moon's surface, researchers can determine the frequency and intensity of past impacts.

Our study of the Moon also contributes to our existing knowledge of planetary atmospheres. While the Moon lacks a substantial atmosphere like Earth's, it does have an exosphere, a very thin layer of gas surrounding its surface. Examining the composition and behaviour of gasses within this exosphere assists scientists in gaining insights into similar phenomena occurring on other celestial bodies throughout the universe.

As lunar science progresses, discoveries continue to be made. For example, the presence of water ice in the permanently shadowed regions near the Moon's poles. This discovery has significant implications for future human exploration and potential colonisation efforts. In addition, the ongoing studies are focused on investigating the possibility of utilising lunar resources for sustainable space exploration missions.

Looking ahead, lunar science holds great promise for uncovering even more mysteries and further expanding our knowledge. By continuing to explore the Moon's rocks, minerals, and elements as well as its atmospheric conditions, scientists not only deepen their understanding of our moon, but other planets and moons within our solar system.

Those lunar samples, brought back by previous missions, have provided scientists with good physical evidence to analyse and study. Scientists employ various analytical techniques to investigate lunar samples. One common method is spectroscopy, which involves studying the interaction of electromagnetic radiation with matter. By analysing the light reflected or emitted by lunar samples, scientists can determine their chemical composition and identify specific minerals and elements present.

It is believed that the Moon holds clues about the early stages of our solar system when it was formed over 4 billion years ago from the debris left behind after a massive collision between a Mars-sized object and the early Earth.

Well, that's one theory.

Studying the Moon's composition can also have practical implications for future space exploration. For example, valuable resources such as water ice or rare metals on the Moon, allow us to assess its potential as a resource-rich destination for future manned missions or even for establishing a lunar base.

So, investigating the Moon's composition is a fundamental aspect of lunar science, but it also has practical implications for future space exploration. Likewise, it is also important to understand more about the Moon's surface, which has resulted from various geological processes that have occurred over its long history.

We also need to understand, not just the Moon's surface, but what lies under it. For example, how deep would the foundations of a man-made structure have to be, and how would the severe temperature fluctuation affect these foundations?

This lunar geology provides valuable information into the formation and evolution of the Moon and may even shed some light on the early history of our solar system.

This view of the north pole region of the Moon was obtained by NASA's Galileo camera during the spacecraft flyby of the Earth-Moon system on December 7 and 8, 1992.

One of the most prominent features seen on the Moon's surface is its craters. These impact craters are formed when asteroids or meteoroids collide with the Moon's surface. The fierce heat generated by these impacts melts the rocks and creates circular depressions. Over time, subsequent impacts, and signs of minor erosion have shaped these craters into what they are today.

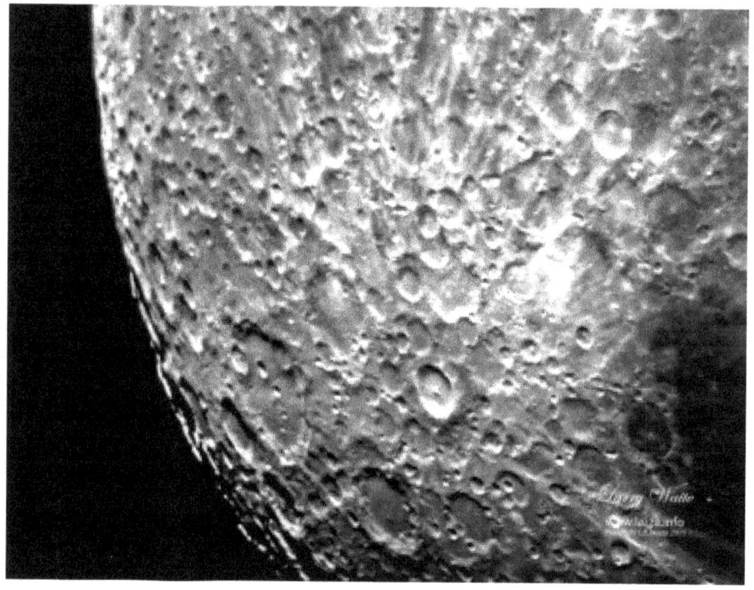

Photograph by the Author

Volcanic activity also played a significant role in shaping the Moon's surface. The Moon's volcanic features include vast plains (maria), formed by ancient lava flows. These dark- regions, visible from Earth, are volcanic remnants from when the Moon was still geologically active.

Volcanic eruptions also created volcanic domes and lava channels on the Moon's surface.

Tectonic forces also influenced the Moon's geological history. These forces caused the Moon's crust to crack creating rilles (*faults and fractures*).

Rilles can be found all over the Moon's surface, ranging from small, narrow channels to large, winding valleys. They provide evidence of past tectonic activity and give scientists insights into the Moon's internal processes.

In analysing samples from missions such as Apollo, scientists have determined the age and composition of lunar rocks.

This information allows them to piece together a timeline of events and understand how different geological processes have shaped the Moon and other celestial bodies in our solar system.

Lunar geology certainly offers fascinating insights into the processes that have shaped the Moon's surface throughout its history. Its craters, volcanic features, and tectonic forces hold many mysteries of our closest celestial neighbour, solving those mysteries would enable a better understanding of the formation and evolution of our solar system.

The Moon's exosphere provides a fascinating area of study for lunar science. This exosphere is composed of various gases, although the concentration of these gases is extremely low compared to Earth's atmosphere. The primary component of the Moon's exosphere is helium, which is believed to be derived from the solar wind. The solar wind is a stream of charged particles, mainly protons and electrons, coming from the Sun.

These particles interact with the Moon's surface and release helium atoms, forming the exosphere.

In addition to helium, the Moon's exosphere contains small amounts of other gases, including hydrogen, neon, argon, and various organic compounds. These gases are thought to originate from sources such as volcanic out-gassing, micrometeoroid bombardment, and the degradation of minerals on the lunar surface.

Knowing the Moon's exosphere provides further valuable insights into several areas of scientific interest. One area concerns our understanding of the processes that occur on airless bodies like the Moon.

By studying the composition and behaviour of gases in the exosphere, scientists can gain valuable knowledge about how these gases interact with the lunar surface and escape into space.

Another significant aspect of exploring the Moon's exosphere is its relevance to space exploration and potential future missions. By analysing the distribution and movement of gases within the exosphere, scientists can develop models and strategies for managing and utilising resources on airless bodies.

This knowledge could prove invaluable for future manned missions to the Moon or other celestial bodies where the availability of resources plays a crucial role. In addition, the Moon's exosphere provides valuable insights into similar phenomena occurring elsewhere in our solar system.

As we continue to explore and uncover mysteries about our moon, delving into the complexities of its exosphere adds another fascinating dimension to our understanding of it and its wonders.

Future Moon Missions and a Human Presence on the Moon

Technical data obtained through years of rigorous research has provided several intriguing revelations. These important findings suggest that we are all but ready for more ambitious explorations very soon!

Through our ever-expanding improvements in technology, each passing year we gain a better understanding of not only the Moon but of our solar system, and beyond.

One of the most impactful learnings relating to the Moon was the validation of water ice. Initially presumed to be a parched and desolate place, scientists were able to determine beyond proof that water molecules exist deep in the shadows of some of the polar craters.

This is a real game-changer for future lunar exploration!

This is a fantastic discovery for upcoming missions to the Moon as this water ice can potentially be utilised for human consumption as well as propelling rockets for much deeper explorations.

In addition to water, a significant understanding of the Moon's geological history has been gained through the examination of lunar rocks and some isolated surface elements. This allowed researchers the ability to characterise various geological processes. Processes that have contributed to the Moonscape over a period stretching billions of years.

Apollo 17 -Collecting Samples - AS17-138-21162 - *NASA*

THE TECH FOR FUTURE MISSIONS

Future lunar missions will utilise advanced technologies and instruments that further enhance our understanding of the Moon's surface and composition. One key technology that will play a crucial role is robotics. Robotic rovers, landers, and orbiters will be employed to collect data, capture images, and perform scientific experiments remotely on the lunar surface.

For instance, NASA's Volatiles Investigating Polar Exploration Rover (VIPER) is designed to search for water ice near the Moon's South Pole. Equipped with a drill VIPER will analyse samples for traces of water that could be utilised for future human missions. Another important instrument is spectrometers, which measure the interaction between light and matter to determine elemental composition.

Spectrometers onboard future lunar missions will provide valuable insights into the Moon's geology, mineralogy, and potential resources.

They can help identify areas rich in volatiles such as water ice, and minerals that may have economic value.

In addition, remote sensing technologies will enable high-resolution mapping and imaging of the Moon's surface features. For example, Light Detection and Ranging (LIDAR), uses laser pulses to create detailed 3D maps of the lunar landscape. This technology can aid in identifying safe landing sites, locating resources, and studying geological formations.

There are numerous upcoming lunar missions by various space agencies and private companies that hold promise for expanding our knowledge of the Moon.

As these missions unfold humanity's understanding of our celestial neighbour will undoubtedly deepen bringing us even closer to unlocking the wonders of our moon.

As human exploration of the Moon continues to evolve, there is growing interest in establishing permanent human settlements or bases on its surface. Various space agencies and private companies have put forth plans and proposals for these ambitious endeavours.

One key proposal is the creation of lunar habitats that would provide a sustainable living environment for astronauts. These habitats would require careful design to withstand the harsh lunar conditions, including exposure to extreme temperatures, radiation, and a lack of breathable air and water.

Innovative technologies, such as 3D printing and advanced materials, are being explored to construct any necessary buildings using lunar resources, reducing the need for transporting building materials from Earth.

Another aspect of establishing a sustainable human presence on the Moon involves resource utilisation. The Moon is rich in valuable resources such as the water ice, which can be converted into drinking water, breathable oxygen, and rocket propellant. The ability to extract and utilise these resources would significantly reduce the cost and logistical challenges of lunar missions by minimising reliance on Earth for essential supplies.

There will be an extraordinary number of challenges and risks in the pursuit of establishing human settlements on the Moon. One major challenge is the long-term effects of lunar gravity on human health. Since the Moon has a significantly lower gravity than Earth, prolonged exposure to this reduced gravitational force may have adverse effects on astronauts' bodies, including muscle atrophy and bone density loss.

Additionally, shielding against radiation is crucial for protecting astronauts from the high levels of radiation present on the Moon's surface. Constructing effective shielding systems that can mitigate these risks will be essential for ensuring the safety and well-being of lunar settlers.

Despite these challenges, there are numerous potential benefits to establishing a sustainable human presence on the Moon.

It would serve as an important stepping stone for future deep space exploration missions, providing valuable experience and knowledge for longer-duration space travel. It could also enable scientific research that is uniquely possible on the Moon's surface, ranging from studying lunar geology to conducting astronomical observations without atmospheric interference.

So, the quest to establish a sustainable human presence on the Moon is an exciting frontier in space exploration. Plans and proposals for lunar settlements are being developed with considerations for habitat design, resource utilisation, and addressing challenges such as reduced gravity and radiation exposure.

While many hurdles must be overcome, the potential rewards, technological advancement, and paving the way for future space exploration are immense.

The renewed interest in returning humans to the Moon stems from a vision of pushing the boundaries of exploration and expanding our understanding of the universe. Establishing a sustained human presence on the Moon holds numerous motivations and potential benefits that make it an exciting endeavour.

One of the primary motivations for returning to the Moon is for further scientific research and exploration. The Moon offers a unique environment for studying various aspects of space, including geology, astronomy, and human physiology.

In addition to the awesome scientific opportunities, a lunar base could provide valuable resources for sustaining human life in space.

A base on the Moon could act as a testing ground for technologies and strategies that would be essential for long-duration space flights to destinations like Mars. By mastering the challenges of living on the Moon, we would gain valuable experience and knowledge that could significantly aid future efforts in space colonisation. Overall, the vision of setting foot on the Moon again and establishing a sustained human presence is driven by curiosity, ambition, and the quest for knowledge.

Through these missions, we hope to unlock more secrets of the universe, harness valuable resources, and pave the way for further exploration beyond our celestial neighbour.

The assessment of the different proposed concepts for moon bases reveals the innovative ideas and architectural challenges associated with building structures on the Moon.

One concept for a moon base involves surface habitats that would be constructed on the lunar surface. These habitats would provide living quarters and work spaces for astronauts and researchers. However, surface habitats face several challenges, including protection from radiation and micrometeoroids. The Moon lacks a protective atmosphere like Earth's leaving its surface exposed to harmful solar radiation and impacts from tiny meteoroids. Designing structures that can withstand these hazards is essential for the safety and well-being of those living and working on the Moon.

Another concept explores underground structures as potential moon bases. By building beneath the lunar surface, astronauts could find natural protection from radiation and micrometeoroids. The Moon's regolith, or surface layer of loose rocks and soil, could provide insulation against radiation. Excavating tunnels and incorporating them into the lunar landscape would allow for more efficient use of resources and better thermal stability. However, creating stable underground structures poses engineering challenges that need to be addressed to ensure their structural integrity.

In constructing a lunar base, materials, and technologies suitable for the Moon's unique environment must be utilised. Traditional construction materials used on Earth may not be feasible due to the absence of an atmosphere, extreme temperatures, and low gravity on the Moon. Researchers are exploring alternatives such as lunar regolith bricks made from compacting lunar soil or 3D printing using regolith as a raw material.

These innovative approaches could enable the creation of sturdy and sustainable structures on the lunar surface. Overall, the concept of moon bases brings forth various architectural challenges. Protecting against radiation and micrometeoroids while utilising materials suitable for the lunar environment are key considerations.

Whether through surface habitats or underground structures, designing and constructing lunar bases requires ingenuity and innovation. By addressing these challenges, scientists and engineers are paving the way for a future where humans can establish a sustainable presence on the Moon, unlocking endless possibilities for scientific exploration and advancements in space technology.

Water, Air and Fuel

The water ice on the Moon can be used for several essential purposes. Once the water has been extracted it can be split into its constituent elements, hydrogen, and oxygen, through an electrolysis process.

Hydrogen can serve as a propellant for spacecraft, providing fuel for future lunar missions. Oxygen on the other hand can be used for breathing and as an oxidizer for rocket engines. Therefore, having access to water on the Moon significantly reduces the need to transport these crucial resources from Earth.

In addition to water ice, the iron, aluminium, titanium, and silicon found in the regolith would also be a valuable resource in creating building materials, such as concrete and glass, which would all be considered essential for constructing habitats and other structures on the Moon.

With a few modifications such as adaptations for solar power, hydraulic anchoring systems, and perhaps a type of encapsulation, some existing machinery and technology could be utilised to extract these resources effectively.

For instance, one proposed method involves using solar-powered rovers equipped with drills and excavation tools to collect lunar regolith and retrieve water ice from the polar regions. Once collected, the regolith and ice could undergo processing techniques such as heating, sifting, or chemical reactions to separate desired components.

Management of these resources will certainly play a crucial role in establishing long-term sustainability on the Moon. Developing an efficient recycling system that minimises waste and maximises resource utilisation would be essential.

With state-of-the-art recycling, we could purify and reuse water, regenerate breathable air, and repurpose our waste materials.

International collaboration will also be essential in ensuring responsible resource management on the Moon. Space agencies and private companies from different countries can share expertise, technologies, and knowledge to ensure sustainable practices are implemented.

By working together, we can establish guidelines and protocols that prioritise environmental protection and ethical conduct in lunar resource extraction and utilisation.

The further analysis of lunar resources such as the water ice, and regolith, provides opportunities for sustaining human life on the Moon. By extracting and utilising these resources efficiently we can reduce reliance on Earth for vital supplies like fuel and construction materials.

The Moon's unique environment and proximity to space offer countless opportunities for expanding our understanding of the universe.

Astronomical observations would also thrive on a moon base. Without Earth's light pollution and atmosphere to hinder and distort views, astronomers could peer deeper into space and observe distant galaxies, stars, and phenomena that are otherwise difficult to detect from Earth.

Just imagine a lunar telescope or an array of telescopes that could be established to capture high-resolution images and collect data that could revolutionise our understanding of the cosmos.

Living in a low-gravity environment would also present other intriguing opportunities for research. On the Moon, scientists could study the effects of reduced gravity on human physiology and develop countermeasures for potential long-duration space travel.

By investigating the impact on muscle atrophy, bone density loss, and cardiovascular health, researchers could pave the way for safer and healthier journeys into deep space.

International collaboration and partnerships would play a crucial role in maximising scientific discoveries on the Moon. We could accelerate progress and promote collaborative efforts by pooling resources and expertise from different space agencies worldwide. These efforts could involve joint research projects, data-sharing initiatives, and coordinated experiments that would leverage the capabilities of the various space agencies.

Through these partnerships, we could unlock the full potential of lunar research and make ground-breaking discoveries that could shape our understanding of the universe for generations to come.

To have a base on the Moon would provide extraordinary possibilities. From further studying lunar geology to conducting astronomical observations such as galaxies, stars, and nebulae, there is so much to explore and learn.

The examination of all the challenges and risks associated with long-duration missions, and establishing a human presence on the Moon reveals several key considerations.

Astronauts spending extended periods on the Moon would be exposed to higher radiation levels, which has detrimental effects on their health. Radiation shielding technologies and strategies need to be developed to minimise this risk.

Yet another challenge would be the isolation experienced by astronauts during their time on the Moon. Long periods of isolation can lead to a whole range of health issues. Unlike the International Space Station (ISS), which allows for regular communication with mission control and periodic visits from supply spacecraft, a lunar base would be much more isolated, at least in the early stages.

Astronauts would need to cope with the psychological and emotional effects of extended periods of solitude and limited contact with people back on Earth. Methods for maintaining mental well-being and providing adequate social support would be crucial.

Additionally, the low-gravity environment on the Moon presents its own set of challenges. Prolonged exposure to low gravity can lead to muscle atrophy, bone loss, and cardiovascular changes in astronauts.

To address these issues, exercise equipment and countermeasures would need to be implemented to maintain the physical health of astronauts. It would also be important to understand the long-term effects of lunar gravity on human physiology to effectively mitigate potential health risks.

When contemplating a human presence on the Moon, ethical considerations must come into play.

Responsible conduct in lunar exploration and colonisation efforts is essential for preserving the Moon's delicate ecosystem. Measures should be taken to minimise irreparable environmental damage or impact resulting from human activities and prioritise our sustainable practices.

To overcome these challenges and mitigate these risks, we must continue researching and further develop our understanding of the hazards and associated risks to humans living on the Moon for any period. We must improve radiation shielding materials, and current communication technology, create psychological support systems, learn to live with low gravity and more.

On top of this, we still need to build structures and habitats and establish a viable "Earth-to-Moon" transport system to shuttle people who work and live on the Moon, such as scientists, builders, engineers, and doctors, as well as the ability to resupply the colony.

As outlined above, establishing a sustained human presence on the Moon poses various challenges and hazards that must be carefully considered.

Despite these challenges, an international collaboration could increase our chances of finding solutions to overcome these obstacles and make our presence on the Moon a reality.

AUTHOR'S WRAP UP

I believe the subjects briefly explored in this book are thought-provoking and offer wondrous possibilities and opportunities for mankind.

From determining how the Earth and moon were created, the importance of tides on Earth, and understanding the phases and significance of the Moon for various cultures. For me, the excitement of further exploration, including mankind settling on the Moon one day, is truly mind-boggling.

Several topics in this book of bite-size information contain some simplified explanations at a high level, purposely not diving too deep into the subject matter. However, I offer a handful of questions below for you to reflect on. Undoubtedly, you may have a few questions of your own.

1. If the Moon vanished, what events would occur and how would that affect life on our planet?
2. How do you think the Moon was formed?
3. Something not covered directly: who owns the Moon, and what do you think would happen if a country colonised the Moon and claimed ownership?
4. Does a full moon affect people's behaviour?
5. Where did all the engines from Apollo 11 end up, the ocean, orbit, disintegrated?
6. How would we cope with the extreme weather conditions on the Moon if we lived there?
7. Do you understand the synchronous (tidal lock) rotation of the Moon? Try using two balls: one with an 'X' that represents the near side of the Moon, and rotate them around each other to demonstrate.
8. How many other planets in our solar system have a tidal lock with a moon?
9. Have you ever taken note of the location of the Moon and its phase, at the same time each night?

If this book has sparked a further interest in the Moon for you then I encourage you to dig a little deeper, and the next time you look up at the Moon, remember it holds many secrets yet to be discovered.

Future colonisation

MOONLIT INVITATION

Now that we are nearing the last pages of this book, please don't let the journey end here. Instead, step beyond the confines of written words, venture out into the night, and gaze upon the awaiting silent wonder of our moon.

Set aside a few minutes for an occasional evening, when the world slows down, and the night sky projects its continuous light show on Earth. There, amidst the twinkling stars, the Moon dominates the sky, shifting through its phases. It beckons you to look up, to ponder the mysteries it holds.

Reflect on the tides the Moon creates and their impact on Earth. Ascertain the current phase, download one of many free apps that can depict its rise and setting. Perhaps play some apt music, and share the experience with friends and loved ones.

The Moon is not merely a distant rock orbiting our world, it is our companion, a source of inspiration. They say that gazing at the night moon can relieve anxiety, encourage one to relax and even promote the natural release of melatonin.

There certainly is no harm in looking up at our celestial neighbour, soaking in some of its reflected light, and just taking a special moment to ponder and reflect on life.

I have included a simple "Moon Observation Chart" on page 121 that will allow you to note the Moon phases over a month. If you decide to give it a go, try to make your observations at approximately the same time every night.

ACKNOWLEDGEMENTS

Thanks go to Alison, my publisher, wife, and best friend.

After a few trials and tribulations, I intended to have a single copy of this book printed and gifted to my Wife. This not so much backfired, but evolved somewhat when Alison made the book available to others.

Originally inspired by my recent studies in astronomy, and in particular, lessons about the Moon. Once these studies were complete, I put what I learnt about the various topics into mini books for myself. As I did, I found that there was so much more to learn about the Moon, astronomy, and cosmology. Astronomy and cosmology encompass a vast and awe-inspiring realm of study, delving into the mysteries of the universe on scales both grand and minuscule. From the infinitesimal particles that comprise matter to the unfathomably large structures of galaxies and superclusters, astronomers and cosmologists seek to understand the fundamental principles governing the cosmos.

Cosmologists and Astronomers, explore the birth and evolution of stars, the dynamics of galaxies, the enigmatic nature of dark matter and dark energy, and the origins of the universe itself. Their discoveries not only deepen our comprehension of the universe, but challenge our perceptions of space, time, and reality, revealing a universe of staggering complexity and beauty that stretches far beyond the limits of human imagination. Our moon is just a very tiny piece of this 'astronomical' topic.

Of course, my sincere gratitude and appreciation to NASA for their fantastic work and the use of the various images and information they selfishly make available.

I would also like to comment on Artificial Intelligence (AI). In my quest to harness the power of AI for research and plot ideas, I turned to AI as a trusty digital companion.

While it proved immensely helpful in generating ideas and providing insights, navigating its nuances became a formidable task. Curiously, paragraphs I penned myself were often flagged as AI-generated, leading to a bizarre dance of proving my authorship to sceptical AI detectors. This unforeseen twist often resulted in precious time lost, as I found myself locked in a battle of wits with algorithms designed to distinguish between human and artificial intelligence. I expressed my frustration and asked AI to explain why it gets things wrong and how I tell readers of the fiasco.

I received the above paragraph in reply.

Despite the occasional missteps, however, the journey with AI was an enlightening one, shedding light on the intricate interplay between human creativity and machine intelligence. But I don't think AI is ready yet to replace a human proof-reader.

TERMINOLOGY

Terms used in this book

- **AGGLUTINATE:** Stick together or cause things to stick together

- **ANORTHOSITE**: A light-coloured rock that makes up the Moon's highlands.

- **BASALT**: is an aphanitic extrusive igneous rock formed from the rapid cooling of low-viscosity lava rich in magnesium and iron exposed at or near the surface of a rocky planet or moon. More than 90% of all volcanic rock on Earth is basalt.

- **BRECCIAS**: a rock composed of large angular broken fragments of minerals or rocks cemented together by a fine-grained matrix.

- **DETRITUS:** Disintegrated or worn matter, or debris.

- **EJECTA**: Matter ejected: material thrown.

- **EXOSPHERE**: A thin, atmosphere (like) surrounding a planet or natural satellite.

- **HELIOPHYSICS:** the physics of the Sun and its connection with the Solar System.

- **ISOTOPIC**: One of two or more species of atoms of an element with the same atomic number but different atomic masses.

- **LIGHT YEAR:** The distance light travels in a vacuum in one year (Approximately 9.46 trillion kilometres).
- **MOONQUAKES**: Moonquakes happen on the Moon because of the shifting of tectonic plates. They can be caused by the tidal pull of Earth on the Moon, meteor impacts, or the change in temperature on the surface of the Moon
- **NEUTRONS**: The neutron is a subatomic particle, which has a neutral charge (not positive or negative).
- **OLIVINE**: A group of rock-forming minerals in Earth's crust abundant in Earth's mantle. A constituent of many meteorites.
- **PARALLAX**: This is the observed displacement of an object caused by the change in the observer's point of view.
- **PROTON**: A proton is a stable subatomic particle.
- **PROTOPLANETARY DISK:** A protoplanetary disk is a Rotating circumstellar disk of dense gas surrounding a young newly formed star.
- **PYROXENE:** The pyroxenes (commonly abbreviated Px) are a group of important rock-forming inosilicate minerals found in many igneous and metamorphic rocks.
- **RILLES:** German for 'grooves," rilles are typically used to describe any of the long, narrow depressions in the Moon's surface that resemble channels.
- **REGOLITH:** Regolith is the layer of loose soil and debris

covering the Moon's surface.

- **SIDEREAL:** Sidereal time is a "time scale" based on Earth's rate of rotation measured relative to the fixed stars.
- **SEISMOMETERS:** A seismometer is an instrument that responds to ground noises and shaking such as those caused by quakes, volcanic eruptions, and explosions.
- **SYNODIC**: is the period for a celestial object to rotate once about the star it is orbiting, and is the basis of solar time.

SPACE EXPLORATION INVENTIONS

Some of the many everyday items we now have are thanks to space exploration.

- **ARTIFICIAL LIMBS:** Artificial limbs are the result of NASA's development of absorption materials paired with robotic and extravehicular activities.
- **CAMERA PHONES:** They were developed in the 90's designed small enough to fit into a spacecraft. One-third of all cameras now use this technology.
- **CAT SCANS:** Developed from a need for quality digital images, that in turn, were used to create CAT scanners. (JPL)
- **COMPUTER MOUSE:** Invented to make computer use easier and more interactive.
- **DUST BUSTERS:** Initially created to collect samples on the lunar surface later turned into a cordless miniature vacuum cleaner in 1979.
- **FOIL BLANKETS:** A lightweight material used for the insulation of equipment, space stations and blankets.
- **FREEZE-DRIED FOODS**: From ice- cream to most other foods, the Astronauts had to eat.
- **INFRARED THERMOMETERS:** NASA used to measure deep space infrared energy sources, and later repurposed them for reading a human body's temperatures.

- **JAWS OF LIFE:** Large apparatus used to separate machinery on the Space Shuttle. A small version was later created to free people trapped in vehicles and is widely used today.

- **LAPTOPS:** Man's first portable computer was a by-product of space exploration.

- **SCRATCH-RESISTANT and UV PROTECTIVE LENSES:** Astronauts' helmet visors had to be robust. NASA worked with several manufacturers to develop a solution that is now used in everyday sunglasses.

- **SMOKE DETECTORS:** The original ionization smoke detectors were invented and approved by NASA and Honeywell. They required the ability to accurately detect fires in Skylab.

- **WIRELESS HEADSETS:** Invented to allow astronauts hands-free communications.

- **WATER PURIFIERS:** Developed by NASA to provide astronauts with drinking water.

Lunar Explorations: The Moon in Plain English

LUNAR OBSERVATION CHART

Sunday Date: Time:	Monday Date: Time:	Tuesday Date: Time:	Wednesday Date: Time:	Thursday Date: Time:	Friday Date: Time:	Saturday Date: Time:

REFERENCES

Numerous resources have been used as a learning tool and source of reference. These are the most utilised links for my referencing purposes.

Ancient pages: https://www.ancientpages.com/

Britannica: https://www.britannica.com/

Cambridge Dictionary: https://dictionary.cambridge.org/

Geology: https://geology.com/

International Astronomical Union: https://www.iau.org/

Lunarium: https://www.lunarium.co.uk/[1]

Merriam-Webster Dictionary: MWD[2]

NASA IMAGES: https://images.nasa.gov/

NASA SCIENCE: https://science.nasa.gov/moon/

Planetary.org: https://www.planetary.org/

NOAA: https://oceanservice.noaa.gov/welcome.html

Science ABC: https://www.scienceabc.com/

Space.Com: https://www.space.com/

The Open University: https://www.open.edu/openlearn/

Wikipedia: https://www.wikipedia.org/

1. https://www.lunarium.co.uk/articles/apogee-perigee/

2. http://www.merriam-webster.com/

IMAGE CREDITS

Images by NASA

Page number then details

- **4** AS14-67-9386 (5 Feb. 1971) -—A close-up view of the laser ranging retro reflector (LR3) which the Apollo 14 astronauts deployed on the Moon during their lunar surface extravehicular activity (EVA). While astronauts Alan B. Shepard Jr., commander, and Edgar D. Mitchell, lunar module pilot, descended in the Lunar Module (LM) to explore the Moon, astronaut Stuart A. Roosa, command module pilot, remained with the Command and Service Modules (CSM) in lunar orbit.
- **29** This image of asteroid Vesta from NASA Dawn spacecraft shows a particularly smooth part of the giant asteroid's surface, likely covered in fine-grained regolith material. Date Created:2012-04-12
- **33** Rock sample brought to earth from the Apollo 12 lunar landing mission -NASA ID: S69-60909. S69-60909 (November 1969) -—A close-up view of lunar sample 12,052 under observation in the Manned Spacecraft Center's Lunar Receiving Laboratory (LRL). Astronauts Charles Conrad Jr., and Alan L. Bean collected several rocks and samples of finer lunar matter during their Apollo 12 lunar landing mission extravehicular activity (EVA). This particular sample was picked up during the second spacewalk (EVA) on Nov. 20, 1969. It is a typically fine-grained crystalline rock with a concentration of holes on the left part of the exposed side. These holes are called vesicles and have been identified as gas bubbles formed during the crystallization of the rock. Several glass-lined pits can be seen on the surface of the rock. Date Created:1969-12-04, Center: JSC
- **63** (Newspaper) Origin of Marshall Space Flight Center (MSFC), NASA ID: 9248173 The urgency and importance of the Marshall Space Flight Center's mission in the 1960s was apparent from the beginning. It became even more apparent on April 12, 1962, when the Soviet

cosmonaut, Yuri Gagarin, became the first man in space. Date Created:1962-04-12

- **64** Two Soviet crewmen for ASTP mission photographed at launch pad. NASA ID: S75-32343, (15 July 1975) -—The two Soviet crewmen for the Apollo-Soyuz Test Project mission are photographed at the launch pad at the Baikonur Cosmodrome in Kazakhstan on the morning of the Soviet ASTP liftoff on July 15, 1975. They are cosmonauts Aleksey A. Leonov (left), commander; and Valeriy N. Kubasov, flight engineer. Leonov is waving to well-wishers at the launch pad. The Soviet ASTP launch preceded the American ASTP Apollo liftoff by seven and one-half hours. The American and Soviet spacecraft were docked in Earth orbit for a total of about 47 hours on July 17-19, 1975. PHOTO COURTESY: USSR ACADEMY OF SCIENCES. Date Created:1975-07-15
- **65** Launch Vehicles. NASA ID: 9139569. The Atlas-Agena-4 boosted the Ranger IV spacecraft for the first U.S. lunar impact on April 23, 1962.Date Created:1962-04-23
- **70** AS15-88-11980, scan: JSC, lunar orbit view: Craters Aristarchus & Herodotus
- **73** Schrodinger: Schrodinger Basin Mosaic of Clementine images of the Schrodinger Basin (312 km diameter). In addition to the prominent, dark, cone-shaped feature (white arrow), Schrodinger has an inner ring of mountains partially encircling the basin floor (a 'peak ring complex') and a network of radial and concentric fractures. The projection is polar stereographic, centered on the basin at - 75 ° S, 132 ° E.
- **73** Schrodinger (2) Schrodinger basin on Moon - Image Credit: NASA
- **74** (Tycho) On June 10th, 2011, the Lunar Reconnaissance Orbiter (LRO) captured this striking image of the Tycho crater. The crater is in the southern lunar highlands and is approximately 82 km in diameter. The crater's notable feature is how steep it is which is due to the fact is a relatively young crater (110 million years old).
- **75** KINGS CRATER -As16-122-19580. This vertical view, provided by NASA, shows the King Crater on the lunar surface. It was exposed with

colour-positive film in a hand-held 70mm camera onboard the Command and Service Modules during the Apollo 16 mission's 98th orbit of the Moon. Date Created:1972-04-16 -NASA.

- **76** View of Goclenius and other craters, NASA ID: as08-13-2225. AS08-13-2225 (21-27 Dec. 1968) -This photograph was taken from the Apollo 8 spacecraft with long-focal length lens, looking south at the large crater Goclenius, which is in foreground. Hold picture with Goclenius at bottom center. The three clustered craters are Magelhaens, Magelhaens A, and Colombo A. The crater at the upper right is Gutenberg D. The crater Goclenius is located at 10 degrees south latitude, 45 degrees east longitude, and it is approximately 40 statute miles in diameter. Date Created:1968-12-24 Center: SC
- **90** Astronaut on the Moon surface – *NASA, supplied by Unsplash.*
- **91** Astronaut Harrison Schmitt standing next to boulder during third EVA. AS17-146-22294 (13 Dec. 1972) -—Scientist-astronaut Harrison H. Schmitt is photographed working beside a huge boulder at Station 6 (base of North Massif) during the third Apollo 17 extravehicular activity (EVA) at the Taurus-Littrow landing site. The front portion of the Lunar Roving Vehicle (LRV) is visible on the left. This picture was taken by astronaut Eugene A. Cernan, Apollo 17 commander. Schmitt is the lunar module pilot. While astronauts Cernan and Schmitt descended in the Lunar Module (LM) "Challenger" to explore the Moon, astronaut Ronald E. Evans, command module pilot, remained with the Apollo 17 Command and Service Modules (CSM) in lunar orbit.
- **95** Moon North Pole. NASA ID: PIA00126, this view of the north polar region of the Moon was obtained by NASA's Galileo camera during the spacecraft flyby of the Earth-Moon system on December 7 and 8, 1992. http://photojournal.jpl.nasa.gov/catalog/PIA00126. Date Created:1996-01-29, Center: JPL, Keywords: Moon, Galileo, Secondary Creator Credit: NASA/JPL
- **99** AS17-138-21162, scan: JSC (description not yet available)

Images by the Author

- **01** Full moon, - taken by L. Waite 2023
- **05** Moon with the red target, - L. Waite 2023
- **27** From a Maksutov telescope, - L. Waite 2023
- **32** From a Maksutov telescope, - L. Waite 2023
- **36** Rocky shoreline and Milky way - L. Waite 2018
- **38** Saturn: Maksutov Telescope, - L. Waite 2023
- **43** Waning First Quart (southern hemisphere), - L. Waite
- **62** Night landscape desert-type area, -L. Waite
- **68** Moon craters map, - L. Waite
- **83** "Searching", - L. Waite
- **96** Close-up of moon craters, - L. Waite 2021
- **110** 'Pondering Frog', - L. Waite
- **112** Moon-base', AI-generated

Images supplied by Unsplash

- **03** (Earth rise from the Moon) – supplied by Unsplash
- **16** (Eggs)- supplied by Unsplash
- **17** (Lunar Eclipse) – supplied by Unsplash
- **20** (Rahu) – supplied by Unsplash
- **21** (Birth) supplied by Unsplash
- **25** (Moon's Layers) supplied by Unsplash
- **34** (Ice cream truck at low tide) - supplied by Unsplash
- **40** (Beach at low tide) - -supplied by Unsplash
- **56** (Lovers under the Moon) – supplied by Unsplash
- **58** (Dragon) – supplied by Unsplash

1. http://www.nasa.gov/audience/formedia/features/MP_Photo_Guidelines.html

- **60** (Dragon boat race) – supplied by Unsplash
- **72** (Sea of Tranquility) Lunar module – Unsplash /NASA
- **80** (Astronaut in space) supplied by Unsplash
- **84** (Footprint) Unsplash/Nasa
- **89** (Lunar module) supplied by Unsplash
- **90** (Astronaut on Moon) supplied by Unsplash
- **92** (Two scientists) supplied by Unsplash
- **100** (Robot) supplied by Unsplash
- **107** (Dishes) supplied by Unsplash

Unless otherwise attributed to NASA or Laurence Waite, all other photographs are sourced through Unsplash.com https://unsplash.com/license/

Diagrams

- **13** Apogee and Perigee, L. Waite
- **115** Aphelion and Perihelion, L. Waite
- **26** Moon's mantle, L. Waite
- **35** Tidal forces, L. Waite
- **46** Moon phases, L. Waite
- **121** Moon observation chart, L. Waite
- *Cover photograph and design by L. Waite*

**Alison Waite is the publisher, owner and account holder for aggregator distribution and all other services including management and royalties.*

About the Author

Venturing out to a remote property under the vast Australian sky sparked a transformative journey for me over a decade ago. Captivated by a spontaneous photograph taken one starlit night from around a campfire, the pull of the stars took hold.

From those humble beginnings with a pocket digital camera, my passion for photography and astronomy began to unfold. Driven by an insatiable curiosity, I spent countless nights over several years sometimes travelling hundreds of kilometres each night searching for that perfect shot. That perfect shot for me at the time, was capturing a night landscape or subject in the foreground and our Milky Way in one photograph.

Photography introduced me to the fundamentals and intricacies of capturing light post-and-processing techniques, from learning the three pillars of photography (shutter speed, aperture, and ISO) to the value of a good tripod, a wide-angle lens, and a remote shutter release.

Transitioning into astronomy, telescopes and eyepieces replaced the traditional camera lens and my attention was drawn from earthly subjects to entities within the cosmic expanse.

Now, I can unearth and capture craters on the moon, stars, nebulas, galaxies, and planets. Yet, with each discovery, I'm humbled by the vastness of the universe, realising that the more I learn, the more questions I have.

The convergence of these passions is a time-consuming hobby that gets me out of the house, usually to some dark, lonely, remote location, and for a few hours gives me the sense that the night sky is all mine, yet my existence is almost insignificant.

Today, I continue to chase the clarity of night skies, finding solace in capturing the timeless elegance of nebulas, galaxies, stars, and planets, all whilst usually accompanied by our celestial neighbour.

With over a decade of observing and photographing the night sky, my journey is a testament to the transformative power of curiosity and our affinity with the cosmos. The intricate details of our lunar neighbour are just as fascinating, mysterious, and mesmerising, as all the other celestial bodies.

Over the last few years, I have learnt to appreciate the moon and its importance to us. I often look at and photograph the moon, which makes me wonder even more each time. It is these experiences, that inspired me to create this book and share my admiration.

About the Publisher

In early 2024, my husband of over 35 years, Larry, completed this small book and after a few frustrating complications, graciously presented it to me as a gift. Initially intended solely for personal enjoyment, the book underwent some thoughtful revisions leading to a reduction of pages to become, what we believe, a good, quick informative read.

With my husband's full permission and support, I published his book through the support of an online aggregator. *-Alison Waite*

www.ingramcontent.com/pod-product-compliance
Ingram Content Group UK Ltd.
Pitfield, Milton Keynes, MK11 3LW, UK
UKHW021658190726
13853UKWH00001B/332

9 798227 361424